Active tectonics in the Upper Rhine Graben

Integration of paleoseismology, geomorphology and geomechanical modeling

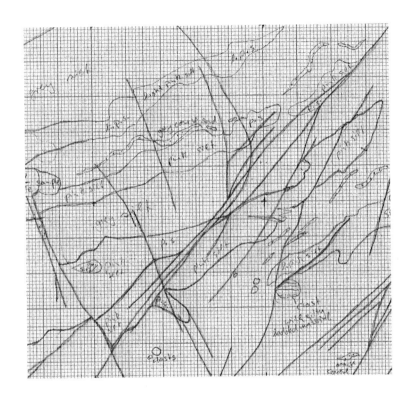

Gwendolyn Peters

Bibliografische Information der Deutschen Nationalbibliothek

Die Deutsche Nationalbibliothek verzeichnet diese Publikation in der
Deutschen Nationalbibliografie; detaillierte bibliografische Daten sind
im Internet über http://dnb.d-nb.de abrufbar.

ISBN 978-3-8325-1650-5

Logos Verlag Berlin
Comeniushof, Gubener Str. 47,
10243 Berlin
Tel.: +49 030 42 85 10 90
Fax: +49 030 42 85 10 92
INTERNET: http://www.logos-verlag.de

VRIJE UNIVERSITEIT

ACTIVE TECTONICS IN THE UPPER RHINE GRABEN
INTEGRATION OF PALEOSEISMOLOGY, GEOMORPHOLOGY
AND GEOMECHANICAL MODELING

ACADEMISCH PROEFSCHRIFT

ter verkrijging van de graad Doctor aan
de Vrije Universiteit Amsterdam,
op gezag van de rector magnificus
prof.dr. L.M. Bouter,
in het openbaar te verdedigen
ten overstaan van de promotiecommissie
van de faculteit der Aard- en Levenswetenschappen
op maandag 10 september 2007 om 15.45 uur
in het auditorium van de universiteit,
De Boelelaan 1105

door

Gwendolyn Peters

geboren te Hamburg, Duitsland

promotoren: prof.dr. S.A.P.L. Cloetingh

prof.dr. F. Wenzel

copromotor: dr. R.T. van Balen

The research reported in this thesis was carried out at the

Tectonics/Structural Geology Department and	Geophysical Institute,
Paleoclimatology and Geomorphology	University of Karlsruhe
Department	Hertzstrasse 16
Faculty of Earth and Life Sciences	76187 Karlsruhe
Vrije Universiteit	Germany
De Boelelaan 1085	
1081 HV Amsterdam	
the Netherlands	

Financial support was provided by the Gesellschaft für Anlagen- und Reaktorsicherheit mbH (GRS, project 150 1252), DAAD (Deutscher Akademischer Austauschdienst, PhD grant), ISES (Netherlands Research Centre for Integrated Solid Earth Science, Rhine Graben project 6.1.5.), Landesstiftung Baden-Württemberg (project 591) and DFG (Collaborative Research Centre/Sonderforschungsbereich 461, project A6).

CONTENTS

CONTENTS...V

ACKNOWLEDGEMENTS.. XI

SUMMARY..XV

SAMENVATTING (SUMMARY IN DUTCH) XIX

ZUSAMMENFASSUNG (SUMMARY IN GERMAN) XXIII

CHAPTER 1

GEODYNAMIC SETTING OF THE UPPER RHINE GRABEN, INCLUDING
THESIS LAYOUT 1

   1.1. Introduction...1

   1.2. Thesis layout..3

      1.2.1. Chapter 1: Geodynamic setting of the Upper Rhine Graben 3

      1.2.2. Chapter 2: Paleoseismology of the Western Border Fault..............................3

      1.2.3. Chapter 3: Pleistocene tectonics inferred from fluvial terraces of the
          northern Upper Rhine Graben 4

      1.2.4. Chapter 4: Tectonic geomorphology of the northern Upper Rhine Graben......4

      1.2.5. Chapter 5: Fault reactivation analysis (slip tendency) of Upper Rhine
          Graben faults..5

      1.2.6. Chapter 6: Synthesis...5

   1.3. Geodynamic evolution of the Upper Rhine Graben...6

      1.3.1. Overview of the Upper Rhine Graben evolution 6

      1.3.2. Prerift setting of the Upper Rhine Graben 7

      1.3.3. Synrift evolution of the Upper Rhine Graben 8

   1.4. Structural setting and faulting mechanisms of the Upper Rhine Graben..............13

      1.4.1. Overview..13

      1.4.2. Major faults of the northern Upper Rhine Graben...15

   1.5. Seismicity of the Upper Rhine Graben and surrounding areas...........................17

      1.5.1. Overview..17

1.5.2. Characteristics of Upper Rhine Graben seismicity 24

1.5.2.1. Source zones of the Upper Rhine Graben 24

1.5.2.2. Magnitude-frequency relationship 25

1.6. The contemporary stress field of the Upper Rhine Graben 27

1.6.1. Stress field determination from an earthquake moment tensor 27

1.6.2. Stress inversion from Upper Rhine Graben earthquake data 28

1.6.3. Present-day kinematics of the Upper Rhine Graben 31

CHAPTER 2

PALEOSEISMOLOGY OF THE WESTERN BORDER FAULT IN THE

NORTHERN UPPER RHINE GRABEN 33

2.1. Introduction 33

2.2. Geology of the Mainz Basin and the northern Upper Rhine Graben 37

2.2.1. The Mainz Basin 37

2.2.2. The northern Upper Rhine Graben 38

2.2.3. The Western Border Fault (WBF) 38

2.2.4. The WBF scarp 39

2.3. Pre-trenching surveys 41

2.3.1. Description of geophysical profiles 44

2.3.1.1. High-resolution reflection seismics 44

2.3.1.2. Geoelectrical tomography profiles 45

2.3.1.3. Ground-penetrating radar (GPR) 47

2.3.1.4. Percussion cores 47

2.3.2. Trenching locations 47

2.4. Trenching descriptions 48

2.4.1. Trench 1 48

2.4.1.1. Overview 48

2.4.1.2. Faulting 50

2.4.1.3. Liquefaction features 54

2.4.1.4. Mass movement features 55

2.4.2. Trench 2 56

2.4.2.1. Overview 56

2.4.2.2. Faulting 56

2.4.2.3. Sedimentary structures 56

2.4.3. Trench 3 57

2.4.3.1. Overview 57

2.4.3.2. Faulting 57

2.4.3.3. Sedimentary structures 59

2.4.4. Trench 4 60

2.4.4.1. Overview 60

2.4.4.2. Faulting 60

2.4.4.3. Sedimentary structures 60

2.5. Structural analysis and timing of the faulting 62

2.5.1. Extensional displacements 62

2.5.2. Timing of the faulting event 62

2.5.3. Structural data 63

2.5.4. Surface trace of the fault system 65

2.6. Reconstruction of trenches 1, 2 and 3 66

2.7. Discussion 71

2.7.1. Discussion on the origin of faults 71

2.7.2. Paleoseismological implications of this study 72

2.7.2.1. Recurrence intervals obtained from paleoseismology and seismology 73

2.7.3. Discussion on the origin of the scarp 74

2.8. Conclusions 76

CHAPTER 3

PLEISTOCENE TECTONICS INFERRED FROM FLUVIAL TERRACES OF
THE NORTHERN UPPER RHINE GRABEN 77

3.1. Introduction 77

3.1.1. Main tectonic features in the region of Pleistocene fluvial terraces 80

3.2. Review of terrace studies 84

3.2.1. Overview 84

3.2.2. Vorderpfalz 86

3.2.3. River Pfrimm 86

3.2.4. Rivers Selz and Wiesbach in the Mainz Basin 87

3.2.5. Lower Main Valley 88

3.2.6. Mainz Bingen Graben 88

3.2.6.1. Mosbach sands / T6 terrace 89

3.2.7. Upper Middle Rhine Valley 90

3.3. New terrace mapping 91

3.3.1. Field mapping 91

3.3.2. Mapping using topographic data 93

3.4. Correlation of terraces 99

3.4.1. Correlation of new terraces 99

3.4.2. Correlation between Mainz Bingen Graben and Upper Middle Rhine

Valley 100

3.5. Fault movements 101

3.5.1. Longitudinal profile 101

3.5.2. Fault movements of the HTBF 109

3.6. Conclusions 113

CHAPTER 4

TECTONIC GEOMORPHOLOGY OF THE NORTHERN UPPER RHINE

GRABEN 117

4.1. Introduction 117

4.2. Regional setting 120

4.2.1. Present-day drainage system 120

4.2.2. Geological setting 122

4.2.3. Fault system 124

4.2.4. Summary of the Rhine system since Late Miocene in context of northern

Upper Rhine Graben tectonics 127

4.3. Investigation of tectonic influence on morphology and drainage 133

4.3.1. Morphological expression of faults 133

4.3.2. Structural control on the drainage system 135

4.4. Quantitative measurements of geomorphic indices 138

4.4.1. Method 138

4.4.1.1. Indices of stream gradient changes 139

4.4.1.2. Mountain-front sinuosity index 141

4.4.1.3. Valley shape index 142

4.5. Geomorphic indices calculations for the northern Upper Rhine Graben 143

4.6. Discussion 149

4.6.1. Interpretation of geomorphic indices 149

4.6.2. Interpretation of structural control on the landscape 152

4.7. Conclusions 155

CHAPTER 5

FAULT REACTIVATION ANALYSIS (SLIP TENDENCY) OF UPPER RHINE

GRABEN FAULTS 157

5.1. Introduction 157

5.1.1. Contemporary stress field and kinematics of the Upper Rhine Graben 160

5.1.2. Crustal structure of the Upper Rhine Graben area 163

5.1.3. Active faults in the Upper Rhine Graben area 164

5.2. Finite element modeling of fault reactivation 167

5.2.1. Introduction to finite element modeling 167

5.2.2. Theory of fault reactivation 168

5.2.3. Theory of slip tendency 170

5.3. Previous studies on fault reactivation 172

5.4. Modeling approach of 3D stress field modeling 174

5.4.1. Geometries of the 3D stress field models 176

5.4.2. Loading of the 3D stress field models 179

5.4.2.1. Model 1 179

5.4.2.2. Problem with the loading of model 1 182

5.4.2.3. Model 2 186

5.5. Results of 3D stress field modeling 192

5.5.1. Stress regime 192

5.5.2. Stress magnitudes and differential stress 195

5.5.3. Summary and comparison of models 1 and 2 198

5.6. Modeling approach for the 3D fault model 199

5.6.1. Geometry of the 3D fault model 199

5.6.2. Separation of fault sets 201

5.7. Slip tendency analysis of Upper Rhine Graben faults 202

5.7.1. Introduction 202

5.7.2. Slip tendency results 203

5.7.2.1. Slip tendency variations with depth 203

5.7.2.2. Slip tendency for all faults and individual fault sets 205

5.7.2.3. Comparison of slip tendency results with independent studies 207

5.8. Discussion 214

5.8.1. Evaluation of stress field modeling results 214

5.8.2. Slip tendency for documented active faults 215

5.8.3. Slip tendency results and relation to regional seismicity 216

5.9. Conclusions 218

CHAPTER 6

SYNTHESIS 221

APPENDIX 227

A.1. Earthquakes in the Upper Rhine Graben and Lower Rhine Graben 227

A.2. Geophysical measurements for pre-trenching surveys 229

   A.2.1. Karlsruhe and vicinity 229

   A.2.2. Forst 230

   A.2.3. WBF scarp 233

      A.2. 3.1. Site 3 – Alsheim 234

      A.2.3.2. Site 4 – Mettenheim 237

      A.2.3.3. Site 5 – Osthofen, north 238

      A.2.3.4. Site 6 – Osthofen, south 240

A.3. Stratigraphy of the trenches 243

A.4. Stratigraphy of terraces in the northern Upper Rhine Graben and Middle
     Rhine Valley 246

REFERENCES 251

ACKNOWLEDGEMENTS

In the first place, I would like to thank my promoter Prof. Sierd Cloetingh for his constant support during the time I worked on the thesis. I especially thank him for the opportunity to perfom the work both at the Vrije Universiteit and the Karlsruhe University and to defend the thesis in Amsterdam, for the many opportunities he gave me to present my work, for his encouragement to publish the individual parts and last but not least for the many constructive discussions.

Next, I would like to thank my promoter Prof. Friedemann Wenzel for the opportunity to perform this research at the University of Karlsruhe, where I have always experienced a very pleasant work environment, for his support and trust in me to manage the research and administrations on my own and for the many fruitful discussions.

Besides my two promoters, I have worked in close cooperation with my two co-promotors. In the first place, I would like to thank Dr. Ronald van Balen for his enthousiasm and interest in my work, for his patience and support during writing and documenting of results (in the end we published 3 papers together!), for his constant support, for taking time for the many constructive discussions and finally for his pleasant company.

Dr. Peter Connolly has been promoting my work in Karlsruhe. I especially thank him for his support and enthousiasm, even for soft and squichy deformations, for his stimulating, sometimes crazy ideas and for his patience with my English, which I think improved during the time I worked on the thesis. I had a great time and learned a lot during several of his friendly field trips and last but not least, we had a lot of fun during numerous Ginsterweg parties and various "projects" (trench roofing, wine making, etc.).

My special thanks go to the members of the reading committee, namely Gabor Bada, Fred Beekman, Bernhard Dost, Kees Kasse, Klaus Reicherter and Peter Ziegler. Their thorough revision and the many fruitful comments helped me to substantially improve the manuscript and to finalize the thesis to become this book. I also thank Meindert van den Berg for taking part in the defense committee.

During my PhD, I worked mostly at the Geophysical Institute at the University of Karlsruhe. I feel very lucky to have been part of the Tectonic Stressgroup during the last years, because working in that group has always been a pleasure. Therefore, I would like to thank all the current and former members of the Stressgroup for the discussions and support, but more importantly, I would like to thank them for the great BBQs, funny lunch talks, cocktail parties and the many other social activities. So all my thanks go to: Johannes Altmann, Mathias Bach, Andreas Barth, Thies Buchmann, Peter Connolly, Jose Dirkzwager, Andreas Eckert, Philipp Fleckenstein, Karl Fuchs, Verena Haid, Oliver Heidbach, Tobias Hergert, Daniel Kurfess, Paola Ledermann, Birgit Müller, Kathrin Plenkers, Gunda Reuschke, Angela Sachse, Eliakim Schünemann, Blanka Sperner, Mark Tingay, Andreas Wüstefeld and Alik Ismail-Zadeh. I thank John Reinecker in particular for his contribution to the trenching project and for his thorough and very helpful reviews of several chapters of this thesis. My special thanks go to Jose for the wonderful time we had in Karlsruhe (there is hopefully more to come in Houston), for her down to earth view, especially when our men again had their heads in the clouds, for her help and for her great company and partner for lunch and womens chats.

The paleoseismology study of this thesis required a lot of organization for the fieldwork campaigns. The members of the E-lab at the Geophysical Institute, namely Manfred Rittershofer, Werner Scherer and Hartmut Thomas, were always helpful providing the buses, tools and geophysical measurement tools.

In addition to the organization of fieldwork, managing the finances of the paleoseismology project was a challenging task, which I could have only performed thanks to the help of Monika Hebben. She was very patient with me, a great company during hours of number crunching and always up for challenging excursions into the power of women...

I have visited Amsterdam many times during the last years. I didn't have my own flat there, but thanks to Magdala Tesauro, Karen Leever and Sandrine Conan I always found a place to stay. I thank all of you for your support and making my visits to Amsterdam always a great pleasure. Magdala, hope to see you around in Europe! And success with your strength map. I hope you don't also have to include the Moon in your model. Karen, success with your thesis and good luck to you in Oslo! And many thanks for always being a helping hand! I have really enjoyed being your roommate.

It was always very enjoyable working at the VU. I would like to thank the members of the Tecto group and the Geomorphology group for their help and support in finding my way around at the VU, for the discussions and the pleasant times we spent during lunches or borrels. Namely, I would like to thank Giovanni Bertotti, Andrei Bocin, Paul Bogaard, Frek Busschers, Maarten Corver, Tristan Cornu, Daniel Garcia-Castellano, Charles Gumiaux, Nico Hardenbol, Ron Kaandorp, Liviu Matenco, Sandra Mertens, Jos de Moor, Diana Necea, Anna du Pree, Alwien Prinsen, Jeroen Smit, Dimitri Sokoutis, Randell Stephenson, Ernst Willingshofer, Gesa Worum, Reini Zoetemeijer.

Thanks to Friedemann Wenzel and Sierd Cloetingh, I was introduced to the Eucor-Urgent and Topo-Europe projects. I would like to thank in particular the Eucor-Urgent members for the very stimulating discussions, the fruitful discussions and the great times during the yearly workshops.

Michael Weidenfeller and Gunther Wirsing from the geological surveys (Rheinland-Pfalz and Baden-Württemberg, respectively) are kindly thanked for the many discussions on the regional geology, tectonics and sedimentary environment and for providing data.

Sven Hüsges, Hennes Obermeyer, Stefan Reiss and Walter Frei are thanked for their geophysical investigations, the many hours of measuring and coring, and the useful discussions on the data interpretation. I also would like to thank Mustapha Meghraoui for his support in the beginning of the trenching project and for the discussions on the trenches and regional geology. My special thanks go to Heiko and Brigitta Melius from Bermersheim, who substantially helped during the trenching and measurement campaigns with logistics, addresses and contacts to the landowners, and who gave me an excellent introduction into Rheinhessen wines.

In the end, I would like to thank my family, my parents, Heinke Peters and Manfred Peters, and Wolfgang Storch, my second father. They have always been a wonderful support for me.

I cannot really express in words how much I thank Thies. Mostly, I thank him for his love, care and help in every aspect of the thesis work (digging, logging, modeling, writing.....) as well as his patience through the entire time I have spent on this thesis.

SUMMARY

Active tectonics in the Upper Rhine Graben – Integration of paleoseismology, geomorphology and geomechanical modeling

In the focus of this thesis is the northern part of the Upper Rhine Graben (URG). The aim of the thesis is to study the active tectonics of the area during the Quaternary and Present. The URG is characterized by intense human modifications, relatively slow tectonic deformation and low intra-plate seismic activity. In addition, during the Quaternary surface processes have been at a high rate. Therefore, the records of the slow tectonic deformation are relatively poorly preserved and thus difficult to detect. Given this setting, several techniques need to be applied in order to study the active tectonics. This thesis demonstrates that the integration of four techniques, paleoseismological trenching, fluvial terrace mapping, quantitative tectonic geomorphology and tectonic modeling, can lead to positive results. *Chapter 1* of this thesis presents a review of the geodynamic evolution, the structural setting and the seismicity of the URG.

The results of paleoseismological trenching, including shallow geophysical measurements, are presented and discussed in *Chapter 2*. This investigation focuses on the Western Border Fault (WBF) in the northern URG. The trenches exposed near-surface deformation of the border fault in Middle to Late Pleistocene deposits (fluvial sands, alluvial loess). Since unambiguous proof for co-seismic deformation is not possible for this site two scenarios are proposed: a single M_w 6.5 earthquake or aseismic creep at a rate of ≥ 0.04 mm/yr. Using 14C and thermoluminescence dating the last faulting episode of the WBF at this site was dated between 8 and 19 ka. Based on a reconstruction of the sequence of events, the interplay between tectonic activity of the WBF and fluvial and erosional processes at the trench site could be demonstrated.

At a regional scale, the effects of fault activity on fluvial terraces are investigated (*Chapter 3*). The work includes morphological terrace mapping along the WBF segment investigated by trenching. Correlation of these newly mapped terraces with previous terrace mapping along the western side of the northern URG permits construction of a longitudinal profile. Based on this profile it is shown that various faults have displaced terraces of Early to Middle Pleistocene age. At this stage, the correlation and determination of displacements depends mostly on a single dated stratigraphic unit (Mosbach Sands). Additional dating of terrace deposits is urgently needed. However, keeping these uncertainties in mind, this study uses the available data to provide a first order quantification of tectonic movements. The uplift rates in the order of 0.01 – 0.08 mm/yr confirm relatively low deformation rates for this part of the URG. The terrace study demonstrates also that the uplift was discontinuous. A significant pulse of uplift occurred first in the northern URG during the Early Pleistocene and subsequently in the Middle Rhine area during the Middle Pleistocene.

Chapter 4 presents an analysis of the tectonic geomorphology of the northern URG. At first, the paleogeomorphic evolution of the study area is reviewed using information on the records of fluvial terraces, the drainage system, on sediment distributions and fault mapping. On the basis of paleogeomorphic maps, the impact of differential uplift of the western margin on the drainage system and the landscape evolution of the graben and the shoulder areas during Late Miocene to present times is demonstrated. Using an overlay of topographic features, drainage patterns and faults, the interaction of both the morphology and drainage pattern with the fault network is investigated and several fault scarps and tectonically influenced parts of the drainage system are identified. Quantitative measurements of geomorphic indices, calculated for stream profiles of tributary catchments of the River Rhine and for the present-day topography, point to active segments of the border faults. Integration of all results of *Chapter 4* supports the significant impact of regional uplift on the morphological evolution of northern URG and the dynamics of its drainage system. It is concluded that the tectonic morphology now preserved in the northern URG resulted from long-term, low level tectonic activity and has been preserved owing to a decrease in erosive activity during the last 15,000 years.

The fourth component of this thesis addresses the potential for fault reactivation under the present-day stress field (*Chapter 5*). Using finite element (FE) modeling techniques, the 3D stress state of the URG and surrounding areas is simulated and then used as input for a 3D slip tendency (ST) analysis of URG faults. Stress field modeling and ST analysis predict that the URG system is in a stable state. Present-day stress field conditions are characterized by relatively low horizontal stresses and subsequently low differential stresses, as well as hydrostatic pore pressure conditions. It is assumed that faults exhibit typical friction coefficients (μ 0.4 – 0.6). Under these conditions, the URG faults are not likely to be reactivated. On the basis of ST results, this study tries to assess the location of possibly active fault segments in the URG area under present-day stress field conditions. Quaternary activity has been documented for several faults from trenching (*Chapter 2*), geomorphology (*Chapters 3 and 4*) and previous studies. High ST values on these faults, mainly striking 170°, support the high potential for reactivation. However, seismicity is not documented for faults of this orientation. A plausible explanation for the observed displacement on these faults is that they are creeping faults. Few damaging earthquakes are recorded for faults with an average 65° strike. This orientation is not favorable for slip and thus could lead to accumulation of stresses and release by earthquakes rather than continuous creep.

The final chapter (*Chapter 6*) presents a summary of results on active tectonics of the URG from this thesis. A variety of method has been applied in this thesis in order to obtain information on the location of active faults, the faulting history, the effects of faulting on the landscape evolution and the fault reactivation potential. The results of this thesis support previous studies stating that tectonic deformation during the Quaternary has been at a low level. However, its imprint on the landscape is preserved because of decreasing erosional activity since Late Pleistocene. Currently, the URG, in particular its northern part, is subject to stable stress field conditions, and the potential for fault reactivation is low. It is suggested that most of the recent deformation in the area takes place as aseismic creep. Integration of all new data results in an updated active fault map for the URG that is presented at the end of this thesis.

SAMENVATTING (SUMMARY IN DUTCH)

Actieve tektoniek in de Boven Rijndal Slenk – integratie van paleoseismologie, geomorfologie en geomechanische modellering (translated title)

Het doel van dit proefschrift is het onderzoeken van de tektoniek in het noordelijke deel van de Boven Rijndal Slenk (BRS) gedurende het Kwartair. De BRS maakt deel uit van een van de grootste tektonische strukturen in de ondergrond van centraal Europa, het "West Centraal Europeese Rift Systeem". De noordelijke deel van de BRS wordt gekenmerkt door intensieve antropogene veranderingen van het landschap, de relatief geringe tektonische bewegingen van de ondergrond en de lage seismische intensiteit van de noordwest Europeese plaat. Tijdens het gehele Kwartair traden intensieve oppervlakteprocessen op, zoals verwering, erosie en rivierwerking. Het onderzoek naar restanten van informatie over tektonische bewegingen van de Rijndal Slenk is door het samenspel van al deze processen bemoeilijkt. Dit proefschrift onderzoekt de aanwezigheid van aktieve tektoniek in de noordelijke BRS door middel van het combineren van informatie met behulp van 4 verschillende methoden: het graven en bestuderen van paleoseismische onderzoekssleuven (onderzoek naar aardbevingen in de recente geschiedenis), rivierterrassen karteringen, geomorfologie (leer van het landschap) en een tektonische modelleerstudie gebaseerd op de eindige elementen methode.

In *hoofdstuk 1* van dit proefschrift wordt een overzicht gepresenteerd van de geologische ontstaansgeschiedenis van het gebied rond de noordelijke BRS. Dit hoofdstuk gaat ook in op de mate waarin bepaalde strukturen belangrijker zijn of zijn geweest dan andere en in hoeverre er seismiciteit (maat voor het optreden in de tijd van aardbevingen van een bepaalde sterkte voor een bepaald gebied) heeft plaatsgevonden. *Hoofdstuk 2* gaat in op de resultaten van de paleoseismische onderzoekssleuven langs een 20 km lang segment van de westelijke randbreuk in het noordelijk deel van de BRS. Het geofysisch onderzoek voorafgaand aan de onderzoekssleuven wordt ook in dit hoofdstuk gepresenteerd. Het vermoeden dat langs deze randbreuk tektonische aktiviteit waar te nemen zou moeten zijn, is bevestigd door de datering van sedimentmonsters genomen tijdens en na de graafwerkzaamheden. De overeenkomst tussen tektonische aktiviteit langs deze randbreuk en laat-Pleistocene fluviatiele erosieprocessen, zoals het zich verleggen van een rivierloop, worden belicht. Het belangrijkste resultaat van dit hoofdstuk is dat in het algemeen uitgegaan kan worden van een wisselwerking tussen aktieve tektoniek en landschapsontwikkeling.

Daarnaast kan deze wisselwerking met behulp van geomorfologische methoden zichtbaar gemaakt worden. Deze relatie is verder verdiept door het terrassenonderzoek in *hoofdstuk 3*. De top van de rivierafzettingen (terrassen) langs 20 km lange segmenten van de westelijke randbreuk alsmede de omringende zuidelijke gebieden zijn door middel van veldwerk en kaartinterpretatie ingedeeld, vergeleken en gekorreleerd met terrassen uit eerdere karteringen van de noordelijk en zuidelijke omringende gebieden. Op basis van korrelatie van even oude terrassen is een terrassenprofiel gemaakt langs de gehele westelijke randbreuk van het noordelijk deel van de BRS. Met behulp van dit profiel zijn relatieve bewegingen langs breuken en vertikale bewegingen van de rivierterrassen bepaald, welke geleid hebben tot bewegingssnelheden.

In *hoofdstuk 4* wordt een overzicht gegeven van de geomorfologie van de noordelijke BRS. Het doel van de combinatie van de resultaten uit *hoofdstuk 3* (het riviernetwerk en de diktes van de verschillende sedimentpakketten) en dit hoofdstuk is een kaart van aktieve breuken in de BRS te realiseren. Kwantitatieve metingen van het landschap zijn gebruikt om de invloed van het gesteente en/of de tektoniek op de morfologie en het riviernetwerk te bepalen en zo de balans tussen erosie en tektoniek op te maken.

In het *vijfde deel* van dit proefschrift wordt de mogelijkheid van het weer aktief worden van breuken onderzocht met behulp van de eindige elementen methode. Een drie dimensionaal model, gebaseerd op de "slip tendency analyse" van Morris et al. (1996) waarin alle reeds gekarteerde breuken alsook die breuken die in dit proefschrift als belangrijk naar voren komen, wordt gebruikt om de huidige spanningstoestand van de noordelijke BRS na te bootsen. Deze resultaten geven een relatief stabiel spanningsveld weer waarin de kans op breukreaktivatie zeer gering is.

Hoofdstuk 6 tenslotte, presenteert een synthese van de resultaten van het onderzoek dat uitgevoerd is naar de oorzaak van aktieve tektoniek tijdens het Kwartair in de noordelijke BRS. Samengevat laten de onderzoeken en de gepresenteerde kaart zien dat paleoseismologie, geomorfologie en de huidige spanningstoestand op en rond de breuken, duiden op een relatief geringe Kwartaire tektonische aktiviteit. Ondanks een aanhoudende beweging van 0.02-0.08 mm/jaar zijn de effecten daarvan op het landschap slechts bewaard gebleven door de afname in erosie sinds het laat-Pleistoceen. De simulering van de huidige spanningstoestand van de Rijnslenk laat zien dat de kans op nieuwe breukbewegingen in het noordelijke deel betrekkelijk laag is. Vermoedelijk is de recente deformatie/beweging in de Rijnslenk voornamelijk het gevolg van kruipprocessen van de ondergrond.

ZUSAMMENFASSUNG (SUMMARY IN GERMAN)

Aktive Tektonik des Oberrheingrabens – Integration von Paläoseismologie, Geomorphologie und geomechanischer Modellierung (translated title)

Ziel der vorliegenden Doktorarbeit ist es, die Tektonik des nördlichen Oberrheingrabens (ORG) vom Quartärs bis heute zu untersuchen. Der Oberrheingraben bildet den zentralen Bereich des sogenannten Westeuropäischen Riftsystems, ein System von tertiären Gräben, das bis in die heutige Zeit seismisch aktiv ist. Die Region des ORG ist gekennzeichnet durch intensive anthropogene Veränderungen und schwache Intraplatten-Seismizität. Im Quartär waren Oberflächenprozesse besonders intensiv, während die tektonische Deformation im gleichen Zeitraum relativ gering war. Aufgrund dieser Gegebenheiten sind die Spuren der insgesamt geringen tektonischen Aktivität nur schwach erhalten und daher schwierig zu untersuchen. Die Untersuchung der aktiven Tektonik erfordert daher die Anwendung verschiedenster Methoden. Im Rahmen dieser Arbeit wurden vier Methoden angewandt und kombiniert: paläoseismologische Grabungen, Flussterrassenkartierung, tektonische Geomorphologie und geomechanische Modellierung.

Im *ersten Kapitel* der vorliegenden Arbeit wird ein Überblick gegeben über die geodynamische Entwicklung, das Strukturinventar und die Seismizität der Oberrheingrabenregion. *Kapitel 2* beinhaltet die Darstellung der Ergebnisse paläoseismologischer Grabungen sowie begleitender geophysikalischer Voruntersuchungen. Diese Geländearbeiten konzentrierten sich auf ein 20 km langes Segment der westlichen Grabenrandstörung im nördlichen ORG. Die Grabungen liefern Hinweise für spätpleistozäne Aktivität, 19.000 bis 8.000 Jahre alt, dieses Störungssegments in Form von Abschiebungen, die mit einer Bebenmagnitude von 6,5 assoziiert werden können. Eine eindeutige Zuordnung der geologischen Beobachtung zu einem 6,5 Beben ist jedoch nicht möglich und Kriechbewegungen mit einer Rate von ≥ 0.04 mm/J sind ebenso wahrscheinlich. Des weiteren zeichnet sich die Grabungslokalität von einem Zusammenwirken tektonischer Aktivität der Grabenrandstörung und fluviatilen Erosionsprozessen aus.

Im regionalen Maßstab wurden die Einflüsse tektonischer Aktivität auf die Landschaftsentwicklung mit Hilfe geomorphologischer Methoden aufgezeigt. Hierzu wurden Flussterrassen am Westrand des nördlichen ORG untersucht (*Kapitel 3*). Dieser Teil der Arbeit beinhaltet die Kartierung von Flussterrassen entlang des 20 km langen Segmentes der westlichen Grabenrandstörung sowie in den angrenzenden südlichen Gebieten. Kartiert wurden die Oberflächen der Terrassen durch Geländebeobachtungen und Interpretationen topographischer Karten. Diese erstmals kartierten Terrassen wurden mit Terrassen aus früheren Kartierungen in den angrenzenden nördlichen und südlichen Regionen verglichen und korreliert. Auf Basis einer einheitlichen Terrassenkorrelation konnte ein Terrassenprofil entlang des gesamten westlichen Grabenrandes im nördlichen ORG erstellt werden. Anhand dieses Profils wurden Relativbewegungen an Störungen (Abschiebungen) und Hebungen von Flussterrassen mit Bewegungsraten von 0.01 – 0.08 mm/J für die letzten ~800.000 Jahre ermittelt. Die Hebung im Untersuchungsgebiet verlief zeitlich versetzt. Erste signifikante Hebungen setzten während des frühen Pleistozäns im nördlichen ORG ein, während die Hebung des Mittelrheintales erst ab dem mittleren Pleistozän verstärkt zu beobachten ist.

Kapitel 4 der Doktorarbeit beinhaltet eine Auswertung der tektonischen Geomorphologie des nördlichen ORG. Diese Untersuchungen umfassen die Auswertung von Flussterrassen, des Gewässernetzes, der Verteilung von Sedimentmächtigkeiten und Störungen mit dem Ziel, Hinweise auf aktive Störungen in der Region zu erhalten. Die Untersuchung wird gestützt durch quantitative Messungen der Geländeoberfläche, die Angaben liefern können, über lithologische oder tektonische Steuerungsmechanismen auf die Morphologie und das Gewässernetz. Die Messungen ermöglichen es zudem, das Gleichgewicht von Erosion und Tektonik regional zu bestimmen. Der Vergleich von topographischen Strukturen mit dem Gewässernetz und Störungsmuster zeigt, dass die Orientierung von Flüssen durch tektonische Aktivität gesteuert wurde. Des weiteren wurden zahlreiche tektonische Bruchstufen im Untersuchungsgebiet identifiziert. Die quantitativen Messungen berechnet an Nebenflüssen des Rheins und für die heutige Topographie, liefern Hinweise auf tektonisch aktive Abschnitte der Grabenrandstörungen. Die Integrierung aller Ergebnisse aus *Kapitel 4* stützt die Beobachtung aus der Terrassenanalyse, dass tektonische Hebung im nördlichen ORG einen bedeutenden Einfluss auf die Entwicklung der Landschaft und des Gewässernetzes gehabt hat. Diese tektonisch geprägte Morphologie ist während des Quartärs bei anhaltenden niedrigen

Deformationsraten entstanden. Sie ist bis heute erhalten geblieben, da die Erosionsaktivität der letzten 15.000 Jahre sehr gering war.

Im *5. Kapitel* der Doktorarbeit werden mit Hilfe einer dreidimensionale Gleitflächenanalyse (slip tendency analysis) Störungssegmente im gesamten ORG identifiziert, die unter heutigen Spannungsbedingungen ein hohes Reaktivierungspotential haben. Die Untersuchung wurde an neu identifizierten Störungen aus dieser Doktorarbeit als auch an bereits kartierten Störungen im Gebiet durchgeführt. Für die Berechnung der Gleitflächenanalyse wurde die Finite Elemente Methode genutzt. Zunächst wurde der heutige Spannungszustand des ORG und angrenzender Gebiete in 3D simuliert. Der ermittelte 3D Spannungstensor diente im Folgenden als Eingangsparameter für die Gleitflächenanalyse. Die Auswertung der Spannungsmodellierung und der Gleitflächendaten zeigt einen stabilen Spannungszustand und ein insgesamt geringes Reaktivierungspotential der Störungen im ORG bei Annahme von Reibungskoeffizienten von $\mu = 0.4 - 0.6$. Die Ergebnisse der Gleitflächenanalyse zeigen, dass ein hohes Reaktivierungspotential für $170°$ streichende Störungen vorliegt. Störungen dieser Orientierung sind u.a. die im Quartär aktive westliche Grabenrandstörung (*Kapitel 2*) als auch aktive Störungen im nördlichen ORG, identifiziert aus den Arbeiten von *Kapitel 3* und *4*. Diese Störungen weisen keine Seismizität auf, jedoch erhöhte Versatzbeträge. Eine möglich Erklärung hierfür ist, dass Störungen dieser Orientierung durch Kriechbewegungen aktiviert werden. Im Gegensatz dazu treten Erdbeben in der Region gehäuft an $65°$ streichenden Störungen auf. Diese Störungen weisen ein geringes Reaktivierungspotential auf, da sie nahezu senkrecht zur größten Hauptnormalspannung orientiert sind. Es ist daher möglich, dass an diesen Störungen aufgrund ihrer Orientierung größere Spannungen aufgebaut werden können, die sich dann in messbaren Erdbeben abbauen.

Das abschließende Kapitel (*Kapitel 6*) der Doktorarbeit fasst die wesentlichen Ergebnisse zur aktiven Tektonik des ORG zusammen, die Mithilfe der verschiedensten Methoden erzielt wurden. Zusammengefasst zeigen die Untersuchungen zur Paläoseismologie, Geomorphologie und zum Störungsverhalten unter heutigen Spannungsbedingungen, dass die tektonische Aktivität im Arbeitsgebiet während des Quartärs relativ gering war. Trotz der geringen Deformationsrate von ca. $0.02 - 0.08$ mm/Jahr sind die Auswirkungen der tektonischen Aktivität auf die Landschaftsentwicklung des nördlichen Oberrheingrabens beträchtlich. Aufgrund der reduzierten Erosion seit dem Spätpleistozän sind die Zeugnisse tektonischer Lanschaftselemente erhalten geblieben. Die Simulation des heutigen Spannungszustandes mit Hilfe der Finiten Elemente Methode zeigt, dass der Spannungszustand stabil ist und dass das Reaktivierungspotential von Störungen als gering einzustufen ist. Es wird vermutet, dass die rezente Deformation im Oberrheingrabengebiet überwiegend durch aseismische Kriechbewegungen stattfindet. Im Abschluss der Arbeit wird eine aktualisierte Störungskarte präsentiert, die alle im Quartär und bis heute aktiven Störungen im Oberrheingraben zeigt.

CHAPTER 1

GEODYNAMIC SETTING OF THE UPPER RHINE GRABEN, INCLUDING THESIS LAYOUT

1.1. Introduction

The Upper Rhine Graben (URG) forms the central part of the European Cenozoic rift system, which consists of rift-related sedimentary basins extending from the Gulf of Lyon to the North Sea (e.g. Illies and Fuchs, 1974; Ziegler, 1992; 1994; Prodehl et al., 1995; Fig. 1.1A). The URG is 300 km long and 30 – 40 km wide. It is bounded to the north by the Rhenish Massif and the Vogelsberg volcano and to the south by the frontal thrusts of the Jura Mountains. The Burgundy Transform Zone in the southwest connects the URG with the southern parts of the European Cenozoic rift system, the Bresse and Rhône Grabens (Fig. 1.1B).

The URG has been the target of many seismic and geological investigations most of which have been carried out for hydrocarbon exploration for Oligocene and Miocene reservoirs during the 1960s and 1970s. This work resulted in detailed knowledge on the Tertiary stratigraphy of the graben fill and on intra-graben faulting and enabled reconstruction of the Tertiary evolution of the rift (for a summary see Pflug, 1982). Little emphasis was placed on the young evolution of the graben so that the Pliocene to Quaternary faulting and depositional history are still poorly understood. Several European network projects, established since the 1990s, have focused on the interaction between recent surface processes and crustal deformation as well as seismic hazards in the Rhine rift system: PALEOSIS (PALEOSIS, 2000), EUCOR-URGENT (see homepage) and ENTEC (Cloetingh et al., 2003; Cloetingh and Cornu, 2005; Cloetingh et al., 2006). In the framework of these projects, studies on the neotectonic activity and seismic hazard of the URG have been performed mainly in the seismically active southern part of the graben (Giamboni et al., 2004; Lopes Cardozo, 2004; Ferry et al., 2005; Fracassi et al., 2005). For the less seismically active northern part of the URG, information on the Quaternary tectonic activity exists for few local structures only (Semmel, 1968; Illies and Greiner, 1976; Monninger, 1985; Haimberger et al., 2005; G. Wirsing, pers. comm. 2005).

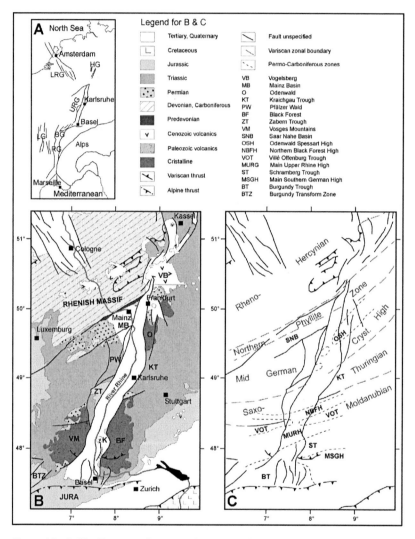

Figure 1.1. A) The European Cenozoic rift system modified after Ziegler (1992) with the centrally located Upper Rhine Graben (URG). LRG = Lower Rhine Graben, HG = Hessian Grabens, BG = Bresse Graben, LG = Limagne Graben, RG = Rhône Graben. B) Geological map of the URG area modified after Lahner and Toloczyki (2004). C) Map showing the extension of the Variscan domains (Rhenohercynian, Northern Phyllite Zone, Mid German

Cristalline High, Saxothuringian, Moldanubian) and Permo-Carboniferous troughs and highs (abbreviations). Map after Boigk and Schöneich (1970) and Franke (1989).

This thesis developed also in the context of the EUCOR-URGENT and ENTEC projects. The aim of the thesis is to study the level and characteristics of recent and past tectonic activity of the URG, particularly of its northern parts, for the period covering the Quaternary to present-day (Fig. 1.1B). The work derives information on a range of topics including URG kinematics, neotectonic activity and the present-day seismic hazard. High rates of Quaternary surface processes, intensive human modification, relatively slow tectonic deformation and presently low intra-plate seismic activity characterize the study area. These characteristics result in many of the features typically associated with recently active tectonic systems (fault scarps, juvenile fluvial systems) being less well preserved, and thus difficult to detect, than in other areas. It is in this context of relatively poor preservation of geological indicators that has required an integrated approach using a variety of different geological and geophysical techniques in order to study successfully the active URG tectonics. This thesis demonstrates the integration of four disciplines often separated in the geological sciences: paleoseismological trenching (*Chapter 2*), fluvial terrace mapping (*Chapter 3*), quantitative tectonic geomorphology (*Chapter 3*) and tectonic modeling (*Chapter 5*).

1.2. Thesis layout

This section summarizes the thesis layout and briefly describes each chapter of the thesis.

1.2.1. Chapter 1: Geodynamic setting of the Upper Rhine Graben

The sections 1.3 – 1.6 of *Chapter 1* give a review of the geodynamic evolution, the structural setting, the present-day seismicity and the stress field of the URG. This review provides background information on the entire URG as well as at the regional and local scale, i.e. the northern part of the URG and the Western Border Fault (WBF) respectively.

1.2.2. Chapter 2: Paleoseismology of the Western Border Fault

The study of paleoseismology presented in *Chapter 2* focuses on the northern

URG, which is characterized by a low level of tectonic activity. Due to this setting, a series of pre-trenching surveys was required to locate an active fault. Using a complete range of shallow geophysical investigations both border faults were investigated at multiple sites. A trench site was chosen at a location where geophysical investigations revealed the best indications for the occurrence of active deformation. This site was located at a segment of the WBF following a 20 km long linear scarp of unclear origin. Four trenches, supplemented with shallow coring, were investigated. The trenching revealed indications for Late Pleistocene deformation, which was dated using 14C and thermoluminescence dating techniques. After a discussion on the origin of this deformation, it is concluded that the origin was tectonic and that the deformation features exposed were associated with activity of the WBF. The paleoseismological implications of this interpretation are also discussed in *Chapter 2*. Additionally, a reconstruction of the sequence of tectonic, erosional and depositional events at the trench site has been built by backstripping. Based on this reconstruction the interplay of activity on the WBF and fluvial and erosional processes is demonstrated.

1.2.3. Chapter 3: Pleistocene tectonics inferred from fluvial terraces of the northern Upper Rhine Graben

The results of the trenching analysis raised a number of questions on the origin of the WBF scarp. The reconstruction of the trench site suggests that the front of the WBF scarp has been terraced by fluvial dynamics under the influence of tectonic activity along the WBF. In order to get a broader understanding of this terrace system and the influence of WBF and other fault activity, a larger region has been investigated further. The area covered the northern URG, the Mainz Basin and the southern part of the Rhenish Massif. For this area, previous terrace studies were compiled and supplemented with new terrace mapping. On the basis of relative height levels, a continuous correlation of terraces from the western margin of the URG to the Rhenish Massif is proposed that permits the study of the transition from the subsiding graben to the uplifted Rhenish Massif. By means of a longitudinal profile the effects of movements along individual faults on the terrace levels and the large-scale regional uplift is demonstrated.

1.2.4. Chapter 4: Tectonic geomorphology of the northern Upper Rhine Graben

This chapter presents the results of a study into the effects of fault activity on the landscape evolution of the northern URG. Information on past tectonic activity is gained from terrace mapping of *Chapter 3*, analyses of drainage directions, sediment distributions, and fault mapping. The compilation of this data is presented as a series of paleogeomorphic maps for Late Miocene to present times. These maps demonstrate the impact of differential uplift of the western margin on the drainage system and the landscape evolution of the graben and the shoulder areas. Recent tectonic effects on the landscape are investigated using an overlay of topographic features, drainage patterns and faults. This allows for investigating the interaction of both the morphology and drainage pattern with the fault network and permits to distinguish fault scarps and tectonically influenced parts of the drainage system. Quantitative measurements of geomorphic indices are used to determine the balance between erosional and tectonic processes. The indices are calculated for stream profiles of tributary catchments of the River Rhine and for the present-day topography. The results point to active segments of the border faults. Integration of all results supports the significant impact of regional uplift on the morphological evolution of the area studied and the dynamics of its drainage system.

1.2.5. Chapter 5: Fault reactivation analysis (slip tendency) of Upper Rhine Graben Faults

The potential for reactivation of URG faults is the focus of *Chapter 5*. Using finite element (FE) modeling techniques, the 3D stress state of the URG and surrounding areas is simulated and then used as input for a 3D slip tendency (ST) analysis of URG faults. Existing fault data and new data obtained from the trenching and geomorphologic studies were incorporated in the 3D slip tendency analysis. The study was performed on faults of the northern URG, as well as on faults of the entire URG and its shoulder areas. Special emphasis in the analysis is placed on those faults with previously documented Quaternary activity and the active faults identified from trenching (*Chapter 2*) and geomorphology (*Chapters 3 and 4*). On the basis of ST results, fault segments that are prone to reactivation under present-day stress field conditions are identified.

1.2.6. Chapter 6: Synthesis

The final chapter presents a summary of results of the work presented in this thesis on active tectonics of the URG. A variety of method has been applied in order to obtain information on location of active faults, faulting history, effects of faulting on landscape evolution and fault reactivation potential. Integration of all new data results in an updated active fault map of the URG, which is presented in the end of the thesis.

1.3.   Geodynamic evolution of the Upper Rhine Graben

1.3.1.   Overview of the Upper Rhine Graben evolution

For a summary of the geodynamic evolution of the entire European Cenozoic rift system the reader is referred to an updated review of Dèzes et al. (2004). The large-scale regional context and lithospheric scale deformation of the Alpine foreland are addressed in the works of Ziegler et al. (1995), Cloetingh and Burov (1996), Cloetingh et al. (1999) and Ziegler et al. (2002).

In the large-scale regional context, the URG and the other rifts of the European Cenozoic rift system developed by passive rifting in the foreland of the Alps. The rifting of the URG was initiated by an E-W extension, occurring approximately contemporaneous with an important phase of the Alpine orogeny (Villemin et al., 1986; Larroque and Laurent, 1988; Dèzes et al., 2004). Its beginning is marked by Middle Eocene synrift sediments revealed from seismic and borehole data (Wittmann, 1955; Sittler and Sonne, 1971; Illies, 1977; Sissingh, 1998). The evolution of the URG occurred in two main phases (Fig. 1.2). During the Oligocene, pre-existing weakness zones were reactivated in an overall extensional to transtensional stress field with E-W oriented extension. Under these conditions, the graben structure opened and sedimentation occurred across the entire URG. The onset of the Miocene was marked by a reorientation of the stress field and a change to sinistral transtension, persisting until Present. During this second phase, which is characterized by a NW-SE oriented compression and NE-SW oriented extension, the URG system was reactivated by sinistral shearing (Illies, 1975; Schumacher, 2002; Michon et al., 2003; Dèzes et al., 2004). During the Early Miocene, sedimentation concentrated on the northern URG segment, whereas the southern and central segments experienced uplift and erosion (Illies, 1974a; Roll, 1979). Since Late Miocene, the sedimentation occurred again in the entire graben structure (Bartz, 1974; Fig. 1.2).

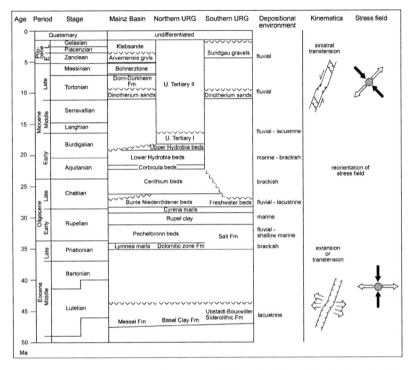

Figure 1.2. Left columns: Lithostratigraphy of the URG and the Mainz Basin based on Sissingh (1998), Grimm (2002) and Grimm et al. (2002). Modified after Schumacher (2002) and Derer (2003). For detailed description of depositional environment see Derer (2003). Right columns: Kinematics and stress field during URG evolution after Illies (1975), Schumacher (2002), Michon et al. (2003) and Dèzes et al. (2004).

1.3.2. Prerift setting of the Upper Rhine Graben

The URG has developed in a region with a complex pre-Tertiary crustal architecture. This had significant influence on the graben geometry. The NNE-SSW trend of the URG is the so-called Rhenish trend, which was established during the Paleozoic Variscan Orogeny. Two other Variscan trends (Erzgebirgian: NE to ENE; Herzynian: NW) are also present in the URG. The Variscan trends are founded in the basement of the URG (e.g. Schumacher, 2002). The basement consists of a sequence of NE trending Variscan highs and troughs separated by fault zones (Fig. 1.1C). Within the URG, the Variscan domains are overlain by NE- to NNE trending Permo-Carboniferous troughs and highs (Fig. 1.1C). A good correlation exists between

Variscan basement structures and the Cenozoic to present-day topography of the graben shoulders. The highest topography of the shoulders is found where Variscan domains are exhumed (Odenwald, Vosges Mountains, Black Forest), whilst Mesozoic troughs and topographic low areas (Kraichgau Trough, Zabern Trough) occur in the Variscan Saxothuringian domain (Fig. 1.1C).

Mesozoic sediments rest discordantly on Permo-Carboniferous series and the basement. In the northern URG, Mesozoic sediments comprise Triassic units while in the southern URG the Mesozoic also comprises Jurassic sediments. Over the entire length of the graben, the Cenozoic graben fill in turn overlies these Mesozoic sediments discordantly. Outcrops of Triassic Buntsandstein along the graben shoulders are found in the northern Black Forest, the northern Vosges Mountains and the Pfälzer Wald. Triassic Muschelkalk is exposed in the Kraichgau Trough and Jurassic sediments are found on the western shoulder in the Zabern Trough and on the eastern shoulder between Freiburg and Basel.

1.3.3. Synrift evolution of the Upper Rhine Graben

The Cenozoic graben fill in the URG consists of a sequence of marine to brackish, fluvial and lacustrine sediments varying in thickness from a few hundred meters to more than 3,000 m (see compilations in Pflug, 1982; Schumacher, 2002; Derer, 2003). In the southern URG, a small depocenter south of the Kaiserstuhl exists. This is filled by 2,500 m of sediments. The northern URG has an elongated depocenter along the eastern graben border with a maximum depth south of Worms of 3,200 m (Doebl and Olbrecht, 1974; Fig. 1.3A).

Synrift deposition in the URG started in Middle Eocene (Lutetian) with terrestrial clays and siltstones in a mainly lacustrine environment (Fig. 1.2). By Late Eocene (Priabonian) two depocenters were established in the URG, the Mulhouse and Strasbourg Basins (Pflug, 1982) the architecture of which is strongly related to the Permo-Carboniferous troughs (Schumacher, 2002). The first major subsidence phase is coupled with a marine transgression during the Early Oligocene (Rupelian) and was associated with an extension of the subsiding area as far north as the Hessian Grabens, northeast of the present-day URG (e.g. Schumacher, 2002). Lacustrine to terrestrial depositional environments dominated until a change to a marine environment culminating in a second marine transgression in the Early Miocene. The location of the depocenter continuously migrated northwards and was eventually established in the northern URG by the Oligocene.

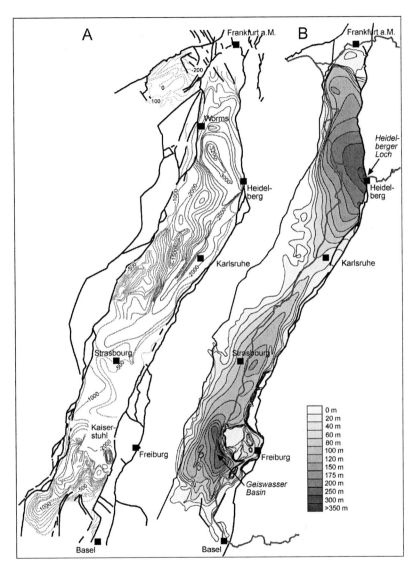

Figure 1.3. A) Depth contours of the base of the Tertiary deposits (relative to the surface) after Doebl and Olbrecht (1974). B) Depth contours of the base of the Quaternary deposits, after Bartz (1974) and Haimberger et al. (2005).

The northern URG and the Mainz Basin are of special interest to this thesis since they were the locus of fieldwork and geomorphological investigations (*Chapters 2, 3 and 4*). For this reason, the geology of this area is presented in more detail in this section. The Mainz Basin is situated at the northwestern end of the URG and was part of the graben in the early stages of rifting from Eocene to Oligocene times. A distinct Mainz Basin existed by Late Oligocene, since when it has evolved independently from the URG. It is important to note that until the Early Miocene (Burdigalian Upper Hydrobia Beds; Fig. 1.2) the stratigraphy is the same for both the URG and the Mainz Basin. Episodic uplift of the Rhenish Massif also involved the Mainz Basin, placing this region in a marginal location of the URG (Fig. 1.3A). The uplift of the Rhenish Massif initiated most likely in Late Cretaceous times and accelerated during Middle Pleistocene (e.g. Van Balen et al., 2000). This is mainly attributed to the isostatic response of the lithosphere to the buoyant effect of the Eifel mantle plume (e.g. Garcia-Castellanos et al., 2000; Ritter et al., 2001).

The Early Miocene was marked by increased tectonic activity in the whole URG area, including the Mainz Basin. The central and southern parts of the graben uplifted, whereas the northern part subsided (e.g. Illies, 1974a). These processes were accompanied by the major volcanic activity of the Kaiserstuhl (e.g. Dèzes et al., 2004) and Vogelsberg (Bogaard and Wörner, 2003). From Middle Burdigalian to Quaternary fluvial and lacustrine sediments with only a few clear stratigraphic markers were deposited in the URG. Hence, the position of the Tertiary-Quaternary boundary, especially in the northern URG, is still under debate (Ellwanger et al., 1995; Fetzer et al., 1995). In the Mainz Basin, a hiatus exists between Middle Burdigalian and Early Tortonian. The onset of Tortonian Dinotherium sands (Bartz, 1936; Weiler, 1952; Grimm, 2002) marks the establishment of a northward oriented River Rhine drainage with source areas lying in the Black Forest and Vosges Mountains (Bartz, 1961; Liniger, 1964). The River Rhine was flowing along the central axis of the URG and across the southern parts of the Mainz Basin (Worms to Bingen; Bartz, 1936; Fig. 1.4) and entered the Rhenish Massif at Bingen. Between Tortonian and Early Pleistocene, the relative uplift of the Mainz Basin forced the River Rhine to migrate northeastward to its present course along the Mainz-Bingen Graben (Wagner, 1930; Kandler, 1970; Abele, 1977; Fig. 1.4). The Mainz-Bingen Graben is located between the Rhenish Massif in the north and the Mainz Basin in the south and is associated with the Hunsrück-Taunus Boundary Fault.

The Quaternary was characterized by continuous fluvial deposition in the URG and by landforming processes with fluvial erosion, loess deposition and solifluction in the Mainz Basin. In the graben, Quaternary deposits show large thickness variations implying differential synsedimentary tectonics (Fig. 1.3B). The deposits concentrated in two depocenters. A southern depocenter, the Geiswasser Basin situated to the southwest of the Kaiserstuhl, filled with 270 m thick sediments and a northern depocenter, the *Heidelberger Loch*, filled with more than 380 m of deposits (Bartz, 1974; Fig. 1.3B). During the Quaternary, the River Rhine and a system of NE-SW tributaries on the western side of the graben have built a flight of terraces in the northern URG. Early and Middle Pleistocene terraces developed in the area referred to as the Vorderpfalz but are at present poorly preserved (terrace mapping by Stäblein, 1968; Monninger, 1985). On the eastern side, continuous subsidence of the *Heidelberger Loch* resulted in the accumulation of fluvial sediments and the absence of terraces. In contrast to the URG, a complete sequence of Early Pleistocene to Holocene terraces of the River Rhine is found at the northern and northwestern part of the Mainz Basin and on the northern side of the Mainz-Bingen Graben (Kandler, 1970; Abele, 1977). The Late Pleistocene lower terrace, formed during the last glacial (Würm), is well defined in the northern URG and the Mainz-Bingen Graben. During the Holocene, the River Rhine incised into the Lower Terrace; periods with more or less pronounced meandering left three distinct meander generations (Fetzer et al., 1995; Dambeck and Thiemeyer, 2002).

The present-day morphology of the URG is characterized by a fluvial plain with the most recent floodplain of the River Rhine situated along the central axis of the graben (Fig. 1.1B). In the northern URG, fluvial incision of tributaries from the western catchment of the River Rhine has resulted in a landscape of ridges and valleys in the area of the Vorderpfalz (Fig. 1.4). These ridges are mainly built up of Pliocene fluvial sands and are covered with Upper Pleistocene loess. The Mainz Basin is at present an elevated area with a tableland landscape of Oligocene to Miocene deposits, which are covered by Pliocene to Pleistocene fluvial sands and Upper Pleistocene loess.

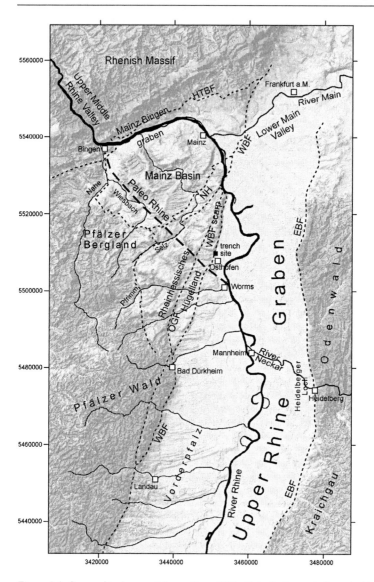

Figure 1.4. Geographical map of the northern URG. Main faults are indicated with dashed lines. EBF = Eastern Border Fault, HTBF = Hunsrück-Taunus Boundary Fault, NH = Niersteiner Horst, OGF = Oppenheim-Grünstadt Fault, WBF = Western Border Fault. Coordinates are in Gauss-Krüger coordinate system. The shaded relief map of this figure and figures 1.8 and 1.10 was created with SRTM data available at http://www2.jpl.nasa.gov/srtm/cbanddataproducts.html.

1.4. Structural setting and faulting mechanisms of the Upper Rhine Graben

1.4.1. Overview

The URG is a highly faulted area. Intra-graben fault mapping is mainly the result of intensive hydrocarbon exploration for Oligocene and Miocene reservoirs at ~ 1 – 2 km depths during the 1960s and 1970s (Fig. 1.5, inset I). Geological mapping at or near surface has identified the faults of the shoulder areas (e.g. Illies, 1974a; Tietze et al., 1979; Stapf, 1988). Unfortunately, little emphasis was placed on mapping the faults in young deposits so that current knowledge of the faults within Pliocene and Quaternary units of the northern URG is poor. Recent reflection seismic investigations on the rivers Rhine, Main and Neckar focused on the structural architecture of Pliocene and younger deposits and provide new insights into younger fault activity particularly in the northern URG (Haimberger et al., 2005; G. Wirsing, pers. comm. 2005).

The kinematics of the intra-graben faults is dominantly extensional although several authors suggest also a few thousand meters of horizontal displacement on them (Bosum and Ullrich, 1970; Meier and Eisbacher, 1991; Ziegler, 1992; Groshong, 1996; Laubscher, 2001). It is important to note that mainly 2D seismic data is available and thus any determination of the horizontal displacement is problematic. The reported vertical displacements of several hundred to thousands of meters (e.g. Pflug, 1982) is much better constrained. Displacements in Quaternary deposits in the order of tens of meters have also been documented (Haimberger et al., 2005; G. Wirsing, pers. comm. 2005).

Figure 1.5 shows fault orientations of intra-graben faults and faults of the shoulder areas. The border faults and intra-graben faults dominantly strike NNE with subsidiary sets striking N to NNW and NW (Fig. 1.5, inset II). Fault strikes in the shoulder areas are more diverse with main orientations of NNE, NE and NW (Fig. 1.5, inset III). The maximum at NNE represents parallel fault strands to the border faults (Rhenish trend). The NE and NW orientations are Variscan fault trends.

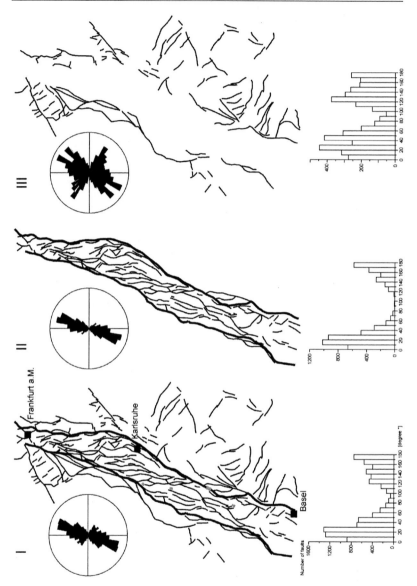

Figure 1.5. Fault map of the URG area and rose plot of strike orientations of faults displayed. Histograms show the frequency distribution of fault strikes. I) Faults of the graben and shoulder areas. Border faults = bold black lines. Active fault strands associated with historical earthquakes marked in grey (after Fracassi et al., 2005; Lopes Cardozo et al., 2005); see section 1.4.1). II) Faults inside URG. III) Faults of shoulder areas. Compilation of fault map after Andres and Schad (1959), Straub (1962), Behnke et al. (1967), Breyer and Dohr (1967),

Illies (1967; 1974a), Tietze et al. (1979), Stapf (1988) and Derer (2003).

Based on 2D seismic profile interpretations, neotectonic fault activity is assigned mainly to 170 – 180° striking faults, which are interpreted by Illies and Greiner (1978) as Riedel shears with a left lateral sense of movement. These fault orientations are not confirmed by interpretations of fault plane solutions, which identified for the northern URG nodal planes dominantly striking 150° and for the southern URG a conjugate set of 20° and 120° striking nodal planes (Bonjer et al., 1984). This observation lead to the hypothesis that second order shear faults are restricted to shallow parts of the upper crust, whereas faults in lower parts of the upper crust strike in direction of the border faults, their conjugate set and with 150° (Bonjer et al., 1984). However, this hypothesis could not be proven yet due to the relatively small number of larger earthquakes in the area for which fault plane solutions could be determined and due to the poor constraint of hypocentral depths and locations of historical earthquakes. Recent seismological studies confirm seismic activity of 20° striking fault strands located in the southern and central URG: the Sierentz Fault west of Basel, the Eastern Border Fault north of Freiburg and small fault strands in the southern and central URG (grey faults in Fig. 1.5, inset I). These faults are associated with important historical earthquakes that, according to Fracassi et al. (2005) and Lopes Cardozo et al. (2005), could be well localized (see also section 1.4.1).

1.4.2. Major faults of the northern Upper Rhine Graben

The Western Border Fault (WBF) strikes in a NNE-direction and bounds the Mainz Basin and the Pfälzer Wald in the northern URG (Fig. 1.6). Two main fault branches run parallel to the WBF and reflect the complex deformation history of the western margin of this part of the graben: the Oppenheim-Grünstadt Fault (after Lampe, 2001); Gundersheim Fault (after Andres, 1958) to the west and the Worms Fault (Ahorner, 1992) to the east, branching from the WBF south of Oppenheim (Fig. 1.6). Along the eastern margin of the northern URG, the Eastern Border Fault (EBF) accommodates most of the vertical displacement, which exceeds 2,200 m along the segment north of Heidelberg. In the Quaternary, tectonic activity remained similar to its Late Tertiary intensity along the EBF resulting in the ~ 380 m deep depocenter of the *Heidelberger Loch* (Fig. 1.3B). The surface trace of the EBF is clearly defined along the foot of the relatively steep mountain-front of the Odenwald. The orientation of the fault is approximately N-S between Frankfurt am Main in the north and Heidelberg in the south. South of Heidelberg, the orientation changes to NNE-SSW. This change in orientation of the EBF and the nearly constant NNE-SSW orientation of the WBF cause a narrowing of the northern part of the graben (Fig. 1.6).

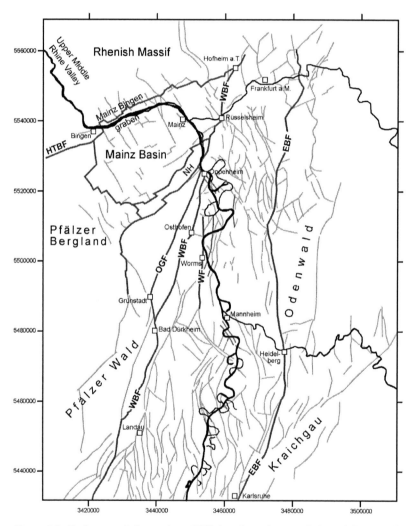

Figure 1.6. Fault map of the northern URG based on the compilation of Figure 1.5. Additionally, faults identified in river seismics have been included (faults north of Mannheim after Haimberger et al., 2005, faults south of Mannheim after G. Wirsing, pers. comm. 2005).

Border faults and major faults are highlighted with thick grey lines. EBF = Eastern Border Fault, HTBF = Hunsrück-Taunus Boundary Fault, NH = Niersteiner Horst, OGF = Oppenheim-Grünstadt Fault, WBF = Western Border Fault, WF = Worms Fault.

In the Mainz Basin, the Niersteiner Horst is a distinct tectonic feature. This structure evolved during Tertiary, contemporary to URG evolution, by reactivation of the Permian Pfälzer anticline. The entire Tertiary sedimentary sequence has been eroded from the horst reflecting an uplift by about 160 m since the Lower Pliocene (Sonne, 1972). Pleistocene uplift of the horst exposed Permian sediments and increased the NW tilting of the sediments (Sonne, 1969).

Significant differences between intra-graben and shoulder fault orientations are also evident in the northern URG (Fig. 1.7). Here, the border faults and intra-graben faults strike dominantly NNE. Subsidiary sets strike N to NNW and NW (Fig. 1.7, inset I). In the shoulder areas three sets of NNE, NE and NW striking faults occur (Rhenish and Variscan trends, Fig. 1.7, inset II). Within the northern URG, geological data from outcrops (Stäblein, 1968; Illies and Greiner, 1976) and reflection seismic profiles (Straub, 1962; Anderle, 1974; Schneider and Schneider, 1975; Derer et al., 2003) give indications for neotectonic faulting mechanisms. The dominant faulting style is of an extensional type and has been identified on all fault sets displayed in Figure 1.7 (inset I).

1.5. Seismicity of the Upper Rhine Graben and surrounding areas

1.5.1. Overview

The seismicity of the URG is characterized by low to moderate intra-plate earthquake activity (Ahorner, 1983; Bonjer, 1997; Leydecker, 2005a). A comparable level of seismicity is observed in the northern segment of the Cenozoic European Rift system, the Lower Rhine Graben (LRG) and the Rhenish Massif (Hinzen, 2003). Both the northern and the central segments of the Cenozoic European Rift system, the latter including the URG, are at present amongst the most seismically active areas in Western Europe north of the Alps (Fig. 1.8). The historical seismic catalogue of the URG area dates back to 800 AD (e.g. Leydecker, 2005a). The map of instrumental and historical earthquakes in the URG shows a wide distribution of small earthquakes occurring over the entire graben (Fig. 1.9). Most remarkable about the seismicity pattern is that the seismicity is not restricted to the border faults of the graben. The WBF is nearly free of seismic activity. Along the EBF, earthquakes occur mainly in the southeastern part of the URG and only at depths below 10 km (Bonjer et al., 1984).

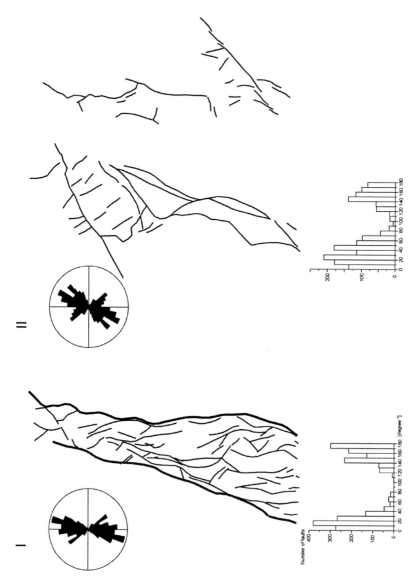

Figure 1.7. Fault map of the northern URG and shoulder areas, rose plot and histogram of

strike orientations of the faults displayed. Based on the compilation of Figure 1.5. I) Faults inside the northern URG. II) Faults of the shoulder areas.

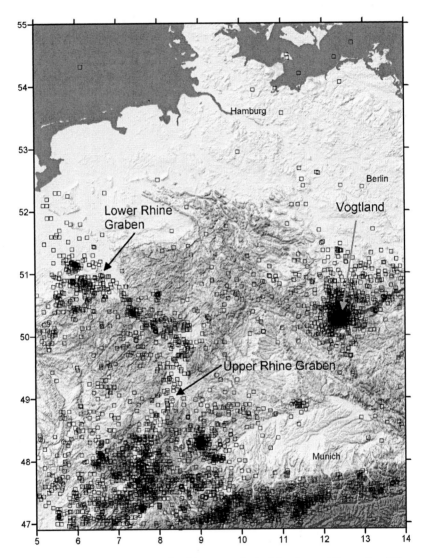

Figure 1.8. Distribution of historical and instrumental seismicity of central Europe from 813 – 2002 AD (maximum intensity I_o = IX; data source: Leydecker, 2005a). Regions of increased seismicity are: Alpine Foreland, Lake of Constance region, Upper Rhine Graben, Swabian Alb, Middle Rhine Graben, Lower Rhine Graben, Vogtland, region of Gera and Leipziger Bucht.

The distribution of earthquakes in the entire URG area shows clear regional variations (Fig. 1.9). The southern part of the graben exhibits a slightly higher activity than the northern part and is characterized by a wide distribution of small earthquakes (M_L 2 – 4) over this part of the graben and its shoulders. Focal depths are mostly between 5 – 15 km within the southern URG and increase to 20 km and greater in the Black Forest (Bonjer et al., 1984; Plenefisch and Bonjer, 1997). This increase of focal depths is also observed south of the URG, in the northern Alpine foreland, where focal depths are typically located in the lower crust (20 – 30 km; Deichmann, 1990; Plenefisch and Bonjer, 1997). A few scattered earthquakes characterize the central part of the URG between Strasbourg and Karlsruhe with focal depths of 10 – 15 km (Bonjer et al., 1984). However, several damaging earthquakes have occurred in historical times in this central segment of the graben (e.g. Rastatt 1933; Seltz, 1952, Fig. 1.10). In the northern part of the graben, small earthquakes ($< M_L$ 4) are also widely distributed but are slightly less frequent than in the south. Focal depths reach a maximum of 20 km, with the majority of events in the upper 10 km (Baier and Wernig, 1983; Bonjer et al., 1984). The main difference between the northern and southern areas is the significantly lower seismicity of the shoulders and adjacent regions in the north, namely the Kraichgau Trough, the southern Odenwald and the Pfälzer Wald. A NW trending belt of increased seismic activity crosses the Rhenish Massif linking the northern end of the URG with the LRG.

It is well documented, but not fully understood, that areas with the highest seismic activity are located at the southern termination of the graben in the Basel area and outside the graben, west of the Vosges Mountains and in the Swabian Alb (e.g. Leydecker, 2005a; Figs. 1.9, 1.10). In these areas earthquakes with magnitudes $M_L > 5$ or intensities $I_o > VII$ have occurred frequently in historic times (Fig. 1.10). The strongest earthquake documented north of the Alps occurred in 1356 at the southeastern end of the URG near the city of Basel. Estimates for the intensity I_o are in the order of IX and for the moment magnitude M_w in the order of 6.0 – 6.6 (Ahorner and Rosenhauer, 1978; Mayer-Rosa and Cadiot, 1979; Mayer-Rosa and Baer, 1992). Paleo-seismological trenching studies suggest that the earthquake reached the threshold of producing a surface rupture and caused rupturing of the Basel-Reinach Fault, south of Basel (Meghraoui et al., 2001). Several independent studies (analysis of damaged stalagmites in caves by Lemeille et al., 1999; slope instability analysis by Becker and Davenport, 2003; identification of liquefaction in lake deposits, Becker et al., 2002, and

Monecke et al., 2004) support the trenching results of Meghraoui et al. (2001), suggesting the recurrence of several earthquakes with estimated magnitudes > 6 in the Basel area. Combining all results, between 5 and 8 M > 6 seismic events have occurred in this area during the past 10,000 years including the event of 1356.

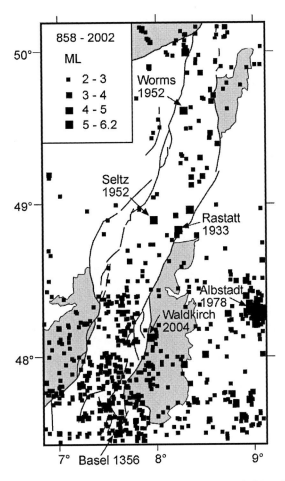

Figure 1.9. Distribution of earthquakes with magnitude M_L > 2 recorded in the URG area between 858 and 2002. For conversion of I_o to M_L the formula $M_L = 0.636*I_o + 0.4$ of Rudloff and Leydecker (2002) has been used. Data source: Fracassi et al. (2005), Leydecker (2005a). The most recent larger earthquake in the URG area reached an intensity of I_o VI, a local magnitude of M_L 5.1 and caused minor and local damage only (Waldkirch i. Br., 05.12.2004). The epicenter was located on the western border of the Black Forest.

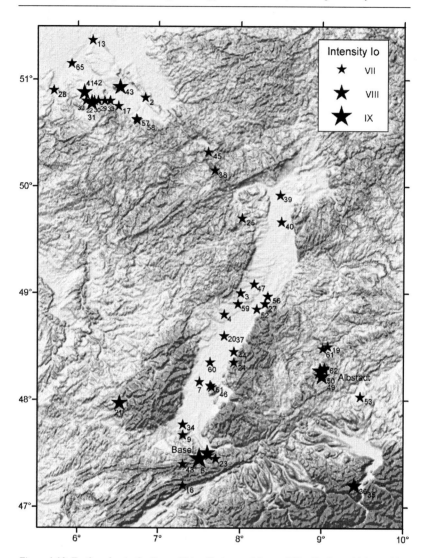

Figure 1.10. Earthquakes in the Upper Rhine Graben and Lower Rhine Graben with intensities I_o > VII from 1021 – 1992 AD (after Fracassi et al., 2005; Leydecker, 2005a; Schwarz et al., 2006). Most of the earthquakes have occurred in historical times and have been damaging. All earthquakes displayed are listed in Table A1.1. The earthquake cluster at Basel includes the earthquakes numbered 1, 5, 7, 8, 12, 14, 15, 18, 23 and 26 in Table A1.1. The cluster at Albstadt includes the earthquakes numbered 49, 50, 54, 55, 61, 62, 63 and 64.

Within the URG, the strongest historical earthquake had intensity VII and an estimated local magnitude of 5.4 (Rastatt 08.02.1933 after Leydecker, 2005a; Fig. 1.9, 1.10). This earthquake was located on a fault sub-parallel to the Eastern Border Fault in the central part of the URG (Ahorner and Schneider, 1974; Fracassi et al., 2005). Few active fault strands within the URG have been identified from the seismicity distribution: the Sierentz Fault west of Basel (Sierentz 1980 earthquake, Lopes Cardozo et al., 2005; the Eastern Border Fault north of Freiburg (Mahlberg/Lahr 1728 earthquake) and small fault strands in the southern and central URG (Kaiserstuhl 1926, Offenburg 1935 and Wissenbourg 1952; see Table A1.1 and grey faults in Fig. 1.5, inset I; Fracassi et al., 2005). The Worms 1952 earthquake is of particular interest to this study since this is one of the few earthquakes whose epicenter was located close to the WBF. This earthquake occurred prior to the setup of a dense seismological network in the area and the determination of the local magnitude and the focal mechanism are of poor quality. A strike-slip mechanism for this event is suggested by Ahorner and Schneider (1974) with a N-S striking nodal plane reactivated sinistrally. Ahorner and Schneider (1974) favor a N-S striking plane since the epicenter was located close to the N-S striking WBF. K. Bonjer (pers. comm., 2004) suggested that an extensional earthquake mechanism might also be possible when combining several earthquakes at the same location to yield a composite fault plane solution.

In contrast to the area of Basel, no recent earthquake in the URG area was large enough to produce a surface rupture. One possible explanation for the lack of large earthquakes could be that the historical time window was too short. In intra-plate settings, large magnitude earthquakes typically have long recurrence intervals of tens of thousands of years. Illies et al. (1981) give an alternative explanation. The authors suggest that the lack of seismicity, in particular along the border faults, is related to the frictional resistance of the fault surfaces. The frictional resistance of these dip-slip faults with dips ranging between 60° and 65° is considered high enough to withstand the shear stresses acting on the fault surfaces. However, this hypothesis does not explain why seismicity is observed on faults in the area, in particular at depths below 10 km. A more conservative view has been proposed by Ahorner (1975) and Fracassi et al. (2005) suggesting that the seismic moment in the URG area is released through smaller events (M < 5). The authors propose that the area is not prone to accumulate sufficient seismic energy for large earthquakes. This interpretation is also supported by Bonjer et al. (1984). They concluded from analysis of P-waves crossing the EBF, that the uppermost 5 – 10 km of the crust are too highly fractured to accumulate sufficient seismic energy. This implies that creep is more likely to occur along the border faults than larger earthquakes.

1.5.2. Characteristics of Upper Rhine Graben seismicity

The characteristics of the seismicity of the URG have been the focus of several studies in the past (Ahorner, 1975; Lippert, 1979; Ahorner, 1983; Bonjer et al., 1984; Ferry et al., 2005). Such seismicity studies have been used for seismic hazard assessments, particularly for the evaluation of site stabilities of nuclear facilities in the URG (Leydecker and Harjes, 1978; G. Leydecker, pers. comm., 2004). The aim of probabilistic seismic hazard assessments (PSHA, based on concept of Cornell, 1968) is to derive earthquake occurrence models that allow prediction of recurrence of large and potentially damaging earthquakes for an area or a specific site. The methodology used in most PSHA consists of four steps. First, earthquake source zones are defined by dividing the area into zones, which are uniform in the character of their seismicity. Second, the seismicity within each source zone is investigated on the basis of the regional earthquake catalogue and parameters such as magnitude-frequency relationships and the seismic rate is determined. Thirdly, site effects and expected ground motions during an earthquake are estimated. Finally, for determination of the hazard at a site, the results from steps 1 to 3 are combined and probabilities for earthquake occurrence at the site are calculated. In the following, the determination of source zones and earthquake recurrence parameters for the URG is presented. In *Chapter 3*, these parameters are compared to recurrence intervals obtained from paleoseismological data gathered as part of this work.

1.5.2.1. Source zones of the Upper Rhine Graben

The URG is commonly divided into two or three zones. For a PSHA, G. Leydecker (pers. comm., 2004) divided the URG in a northern and a southern zone. The division is based on the lower level of seismicity for the northern URG and the higher seismicity of the southern URG. The Basel area, characterized by exceptionally high seismicity compared to the entire URG and surroundings, was defined as a separate zone.

1.5.2.2. Magnitude-frequency relationship

In studies on the characteristics of the seismicity of a study area, the cumulative annual number of earthquakes and their magnitude-frequency relationship are determined using the regional seismic catalogue. The magnitude-frequency relationship follows in most areas the linear Gutenberg-Richter relationship, which describes the size distribution of earthquakes as:

$$\log_{10} N = a - b * M \qquad\qquad Eq. \ 1.1$$

where N is the cumulative number of earthquakes with magnitudes of M or larger and a and b are constants (Ishimoto and Iida, 1939; Gutenberg and Richter, 1944). a is the level of seismicity and describes the number of earthquakes exceeding a specified magnitude for a given area per year. b or "b value" describes the size distribution of events, that is the ratio of small earthquakes to large earthquakes. A higher b value (e.g. b > 1.0) indicates a higher proportion of small events with respect to large events. Since the magnitude and both constants are essential input parameters for PSHA, considerable care must be taken when estimating these parameters. The most critical factor is the b value, which is a statistical measure that is resolved from the gradient of a log-log distribution. Variations in the characteristics of seismicity directly lead to variations in the b value both in space (e.g. Kárník and Klíma, 1993; Wiemer and Wyss, 1997) and time (e.g. Hägele and Wohlenberg, 1970; Urbancic et al., 1992). After Scholz (1968) the b value is dependent on the stress field. Consistently low b values for an area can mean that stresses are high. In addition, rock and fault properties are relevant, so that homogenous rocks and fault asperities can lead to low b values (e.g. Mogi, 1967; Urbancic et al., 1992; Wiemer and Wyss, 1997). Finally, a low b value may also be an indication of an incomplete catalogue. The determination of magnitude-frequency relationships requires a thorough analysis of the earthquake database. A basic assumption for the determination is that earthquake sources are independent. Thus, foreshocks and aftershocks have to be excluded from the database. For the size of the earthquakes, a uniform unit has to be used, which requires conversion of different magnitude types. Large uncertainties in the magnitude determination generally account for historical events, so that a thorough review of the quality of these events is necessary.

Table 1.1 lists magnitude-frequency relationships and the results for recurrence intervals of magnitude 4 to 7 earthquakes derived for the URG, the Basel area and the LRG. It should be noted that most of the authors cited in Table 1.1 do not give errors for their determination of a and b values. On the basis of the seismicity record of the entire URG, the b value is on average -0.73 (Ahorner, 1975; Leydecker and Harjes, 1978; Bonjer et al., 1984). The b value varies significantly if the graben is divided into smaller focal regions (-0.45 > b > -0.94). Lippert (1979) demonstrated that dividing the data set into a northern and southern URG set yields an increase in b values from north to south. The data of Leydecker (pers. comm., 2004) also shows an increase from north to south, although his b values are significantly lower. Lippert (1979) interpreted the relatively lower values for the northern URG to result from the larger number of large earthquakes in this region, which possibly indicated an incomplete catalogue for this region.

Region	a	b	Period	M or I used	M range used	Recurrence interval					Author
						M_L 4	M_L 5	M_L 6	M_L 6.5	M_L 7	
URG	2.2	-0.71	1700-1969	M_L	3-5.2	4	22	115	260	589	Ahorner, 1975
	4.72 ± 0.14	-0.74 ± 0.03	1000-1974	M_L	3.3-6.2	0.02	0.1	0.5	1	3	Leydecker, 1978
	2.24 average value	-0.73 average value	1971-1978	M_L	1.0-2.5	5	26	138	320	741	Bonjer et al., 1984
Northern URG	2.36	-0.58	1971-1978	M_L	0-3.3	1	3	13	26	50	Lippert, 1979
	2.06	-0.74	1971-1978	M_L	0-3.1	8	44	240	562	1,318	Lippert, 1979
	1.23	-0,45	-	I_o	-	20	98	498	1110	2,499	Leydecker, pers. comm., 2004[1]
Southern URG	2.66	-0.94	1971-1978	M_L	0-2.8	13	110	955	2,818	8,318	Lippert, 1979
	2.55	-0.58	-	I_o	-	5	41	328	920	2,617	Leydecker, pers. comm., 2004[1]
	2.8	-0.88	1971-1978	M_L	0-3.6	5	40	302	832	2,291	Lippert, 1979
Basel area	0.86	-0.39	-	I_o	-	23	96	402	914	1,663	Leydecker, pers. comm., 2004[1]
	1.94	-0.79	7,800 years, until 2001	M_w	2-7	9	54	464	1,001	2,929	Ferry et al., 2005, includes paleo-seismic data[2]
LRG	5.85 ± 0.89	-1.0 ± 0.18	1000-1974	M_L	>4	0.01	0,1	1,4	4	14	Leydecker, 1978
	4.64 ± 0.65	-0.85 ± 0.14	1800-1974	M_L	3.8-5.7	0,1	0,4	3	7	19	Leydecker, 1978
Rhenish Massif, LRG	2.44	-0.74	1750-1979	M_L	0-4.7	3	18	100	234	550	Ahorner, 1983
LRG	2.70	-0.90	1700-1999	M_w	4-5.6	8	63	501	1,416	3,981	Ahorner, 2001
LRG, Erft-Sprung	2.06	-0.90	1700-1999	M_w	4-5.6	35	275	2,188	6,166	17,378	Ahorner, 2001
LRG, western faults	1.9 ~ ±0.05	-0.80 ~ ± 0.025	1755-1996	M_w	2-5.6 (6.3 & 6.7)	20	126	794	1,995	5012	Schmedes et al., 2005, includes paleoseismic data
LRG, eastern faults	3.0 ~ ±0.05	-1.20 ~ ± 0.025	1755-1996	M_w	2-5.6	63	1000	15,849	63,096	251,19	Schmedes et al., 2005

Table 1.1: List of magnitude-frequency relationships derived for the URG, the Basel area and the Lower Rhine Graben (LRG). [1] *For conversion of I_o to M_L the formula $M_L = 0.636*I_o + 0.4$ of Rudloff and Leydecker (2002) has been used.* [2] *For conversion of M_w to M_L the formula $M_w = 0.67(\pm 0.11) + 0.56(\pm 0.08)*M_L + 0.046(\pm 0.013) M_L^2$) of Grünthal and Wahlström (2003) has been used.*

Table 1.1 demonstrates also large variations in recurrence intervals. For example, the 1,200-year seismic catalogue of the URG area contains only a single M > 6 earthquake, the Basel 1356 earthquake, thus recurrence intervals of few hundred years for M_L > 6 earthquakes seem unrealistic (Table 1.1). For the Basel area recurrence intervals for Basel 1356 size earthquakes (M_w 6.0 – 6.5) obtained from paleoseismological trenching are in the order of 1,500 – 2,500 years (Meghraoui et al., 2001). For the less seismically active URG recurrence intervals are most likely one order in magnitude higher. Meghraoui et al. (2001) presented a magnitude-frequency distribution for the Basel area supplemented with recurrence intervals of potential Basel 1356 size earthquakes obtained from paleoseismological data. They suggest that for higher magnitude events the distribution might deviate from the linear Gutenberg-Richter distribution towards lower occurrence probabilities. This so-called truncated Gutenberg-Richter distribution is commonly used for regional seismicity studies (e.g. for California, Field et al., 1999; Petersen et al., (2000), and it has also been proposed for the LRG Ahorner, 2001). Contrary to the study of Meghraoui et al. (2001) a more recent study of the seismicity of the Basel area suggests a linear distribution when including both instrumental, historical and paleoseismological data (Ferry et al., 2005).

1.6. The contemporary stress field of the Upper Rhine Graben

The URG stress field has been the subject of several seismological studies (Ahorner et al., 1983; Larroque et al., 1987; Delouis et al., 1993; Plenefisch and Bonjer, 1997; Hinzen, 2003). All of these studies invert stress tensors from earthquake fault plane solutions. The following discussion summarizes the key results of these studies. For a discussion of moment tensor inversion, the reader is referred to Dziewonski et al. (1981), Jost and Hermann (1989) and Stein and Wysession (2003).

1.6.1. Stress field determination from an earthquake moment tensor

A fault plane solution from a seismic event calculates the orientation of the two nodal planes and the orientation of the P-, B- and T-axes. Using the assumption that the three axes correspond to the intra-plate principal stress axes:

P- axis (pressure): maximum principal stress axis (σ_1)

B-axis (neutral): intermediate principal stress axis (σ_2)

T-axis (tension): minimum principal stress axis (σ_3)

it is possible to infer the intra-plate stress orientations from the earthquake mechanism. Following Anderson (1905; 1951) it is commonly assumed that one of the principal stresses is vertical and has a magnitude approximately equal to the weight of the overburden. This stress axis is termed S_v. The two other, mutually perpendicular, principal stresses must then be in the horizontal plane. These stress components are normally termed the maximum horizontal stress (S_H) and the minimum horizontal stress (S_h). Using the three possible relationships between the magnitudes and orientations of the principal stresses, three tectonic "regimes" can be defined (Andersonian stress regimes Anderson, 1905; 1951):

extensional regime $S_v > S_H > S_h$

strike-slip regime $S_H > S_v > S_h$

compressional regime $S_H > S_h > S_v$

1.6.2. Stress inversion from Upper Rhine Graben earthquake data

The calculation of the stress tensor from fault plane solutions has been performed for earthquake data from the entire URG and surrounding areas (Rhenish Massif, Lower Rhine Graben, Swabian Jura and Vosges Mountains; Ahorner et al., 1983; Bonjer et al., 1984; Larroque et al., 1987; Delouis et al., 1993; Plenefisch and Bonjer, 1997; Hinzen, 2003). The quality of the principal stress orientations inferred depends strongly on the quality of the original earthquake data and on the number of earthquake characteristics recorded. For the southern URG and its graben shoulders, a large number of high quality data are available, since earthquakes are frequent and a dense network of seismic stations (German, Swiss and French surveys) exists. This allows precise location and characterization of the earthquakes in the southern URG. For the northern URG and shoulder areas, the earthquake data set is significantly poorer since earthquakes are relatively infrequent and the seismic monitoring network is less dense. The studies of Ahorner et al. (1983), Plenefisch and Bonjer (1997) and Hinzen (2003), therefore, combine data from the northern URG with data from the more seismically active Rhenish Massif and LRG.

The stress inversion studies of Ahorner et al. (1983), Bonjer et al. (1984), Larroque et al. (1987), Delouis et al. (1993), Plenefisch and Bonjer (1997) and Hinzen (2003) point to regional variations of the stress field. Inversions of the principal stresses for the southern URG and the graben shoulders suggest that the region is transtensional, with the strike-slip component being slightly more dominant (Bonjer et al., 1984; Plenefisch and Bonjer, 1997). A NW-SE orientation of S_H is typical for this region. In the northern URG, the Rhenish Massif and LRG the tectonic regime is extensional with a minor strike-slip component. S_H is oriented NW-SE in the northern URG and rotated to WNW-ESE in the Rhenish Massif and LRG (Bonjer et al., 1984; Delouis et al., 1993; Plenefisch and Bonjer, 1997; Hinzen, 2003). Recent modeling of strain rates across the URG and LRG, based on velocity measurements of continuously operating GPS stations, revealed a gradual change from transpression in the southern URG, transtension in the northern URG and extension in the Rhenish Massif and LRG (Tesauro et al., 2005). This study confirms the decrease of compression from south to north, which has been suggested by the seismological studies (Ahorner et al., 1983; Delouis et al., 1993; Plenefisch and Bonjer, 1997). All the studies reviewed state that the URG has no autonomous stress regime, and thus no dominant effect on the present-day stress field.

Comparing the orientations of S_H from south to north, an anticlockwise rotation of ~ 30° for S_H is observed (Ahorner et al., 1983; Bonjer et al., 1984; Delouis et al., 1993; Plenefisch and Bonjer, 1997). Data from borehole breakouts, hydrofractures, overcoring and focal mechanisms released in the World Stress Map (Reinecker et al., 2004) also shows that S_H rotates ~ 15° anticlockwise from the southern URG area (including the shoulders) to the LRG (Müller et al., 1992). Within the URG very few stress data is available. This data indicates predominantly a NW-SE orientation for S_H (Fig. 1.11). At a large scale, the NNE-SSW trending URG is subjected to sinistral transtension under the present-day stress field with northwest directed S_H. With the change in orientation of the Rhine Graben system from NNE-SSW to the NW-SE trending LRG, the tectonic regime changes to extension that is directed nearly orthogonal to S_H (e.g. Ahorner, 1975; Ahorner et al., 1983; Bonjer et al., 1984; Delouis et al., 1993; Plenefisch and Bonjer, 1997; Hinzen, 2003; Tesauro et al., 2005).

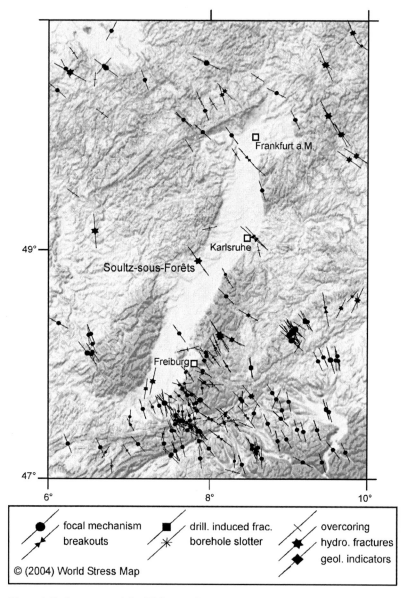

Figure 1.11. Stress map of the URG area showing a NW-SE orientation of S_H. The stress orientations are relatively homogenous and only few stress perturbations occur, most of which lie in the southern URG. Database: World Stress Map, Reinecker et al. (2004).

1.6.3. Present-day kinematics of the Upper Rhine Graben

The present-day stress regimes and the strain rates of the URG area reflect the plate boundary loads (Müller et al., 1992; Plenefisch and Bonjer, 1997). In the large context, the present-day north- to northwest-directed compressional stress regime of Western and Central Europe results from a combination of forces: convergence of Africa-Arabia with Europe resulting in collision in the Alpine orogen and North Atlantic ridge push (e.g. Richardson et al., 1979; Zoback et al., 1989; Grünthal and Stromeyer, 1992; Müller et al., 1992; Richardson, 1992; Gölke and Coblentz, 1996). The ongoing African-Eurasian convergence is well constrained with 3 to 8 mm/year NW to NNW directed motion for the Mediterranean realm (e.g. DeMets et al., 1990; Kreemer and Holt, 2001; Nocquet and Calais, 2004). For the area north of the Alps, current GPS measurements are not yet accurate enough to detect small intraplate motions. Consequently, it remains unclear how much of the Alpine convergence is influencing the present-day kinematics north of the Alps and how much of it is accommodated by the active Alpine mountain belt (Vigny et al., 2002). Based on GPS velocities available at present, the kinematics of intraplate Central Europe area is characterized by a NW push of 1 mm/year and an E-W opening of the ECRIS of 0.5 mm/year (Vigny et al., 2002; Nocquet and Calais, 2004). A recent study of permanent and campaign GPS sites in the URG confirms the results of Vigny et al. (2002) and Nocquet and Calais (2004) with horizontal displacements of less than 0.5 – 1.0 mm/year measured (Rózsa et al., 2005). The short observation time (3 years) only for this study precluded determination of the direction of motion. A modeling study of Tesauro et al. (2005) supports the analyses of the GPS velocities mentioned above. The authors developed a crustal four-block model of Central Europe and calculated strain rates based on the GPS velocities. The modeling yielded a N to NW directed motion of the crustal block located to the east and north of the URG and LRG (0.76 mm/year) and an E-W motion of the crustal block located to the west and south of the URG and LRG respectively (0.51 mm/year).

CHAPTER 2

PALEOSEISMOLOGY OF THE WESTERN BORDER FAULT IN THE NORTHERN UPPER RHINE GRABEN[1]

2.1. Introduction

This chapter presents results of a study on the neotectonic activity of a segment of the Western Border Fault (WBF) in the northern part of the Upper Rhine Graben (URG; Fig. 2.1A). This area is characterized by low seismicity (e.g. Leydecker, 2005a; Fig. 2.1B). Important seismic events were a magnitude 5.1 earthquake (G. Leydecker, pers. comm. 2004) located close to the WBF (Worms 24.02.1952), three intensity VII earthquakes at Mainz (858 AD, 14.03.1733 and 18.05.1733; Kaiser, 1999) and an earthquake swarm in the 19th century. This swarm was located in the central part of the northern graben and produced a maximum intensity of VII (near Gross-Gerau, 1869 – 1871 swarm; Fig. 2.1B). No prominent active faults have been identified from the seismicity distribution alone and no recent earthquake in the entire URG was large enough to produce a surface rupture. On the contrary, the geological record points to a high level of Quaternary tectonic activity of the northern URG. The large thickness variations of Pliocene and Quaternary sediments in this part of the graben imply syndepositional tectonic movements and suggest an average subsidence rate for the Quaternary of 0.1 – 0.2 mm/yr (e.g. Pflug, 1982; Monninger, 1985; Fig. 2.1C). Precise levellings across major faults in the northern URG show contemporary movements of 0.4 – 1 mm/yr (Schwarz, 1974; Demoulin et al., 1995). Several authors use these high present-day movements as evidence for recent tectonic activity (Illies, 1974a; Schwarz, 1974; Demoulin et al., 1995; Franke and Anderle, 2001).

---

[1] This chapter is mainly based on the publication: Peters, G., Buchmann, T.J., Connolly, P., van Balen, R., Wenzel, F., Cloetingh, S.A.P.L., 2005. Interplay between tectonic, fluvial and erosional processes along the Western Border Fault of the northern Upper Rhine Graben, Germany. Tectonophysics 406: 39-66.

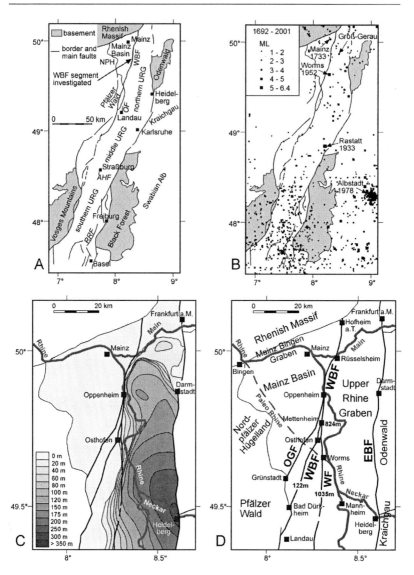

Figure 2.1. A) Structural map of the URG and location of the Western Border Fault (WBF) segment investigated. NPH = Nordpfälzer Hügelland, AHF = Achenheim-Hangenbieten Fault, OF = Omega Fault, RRF = Rhine River Fault. B) Distribution of earthquakes with magnitude $M_L > 1$ recorded between 1692 and 2001 (data source: (Leydecker, 2005a). The Swabian Alb is characterized by the highest recent seismicity, with the Albstadt 1978 earthquake being the largest (M_L 5.6). The Worms 1952 and Rastatt 1933 earthquakes were substantial twentieth

century earthquakes located inside the URG. C) Isopach map of Quaternary deposits (modified after (Bartz, 1974; Haimberger et al., 2005). D) The location of the WBF, the Worms Fault (WF) and the Oppenheim-Grünstadt Fault (OGF) in the northern URG. Offsets along the WBF and WF are measured for the Lower Miocene Hydrobia Beds (Stahmer, 1979). The course of the River Rhine between Upper Miocene and Early Pliocene is indicated with a dashed line.

The most obvious topographic feature of the entire graben is the significant height difference between the graben shoulders and the River Rhine plain (up to ~ 1,000 m), representing a clear morphological signature of the border faults. This suggests that tectonic movements have occurred on the latter during the Quaternary. The River Rhine Fault south of Freiburg, the Achenheim-Hangenbieten Fault near Strasbourg and the Omega Fault system near Edenkoben/Landau are documented faults in the URG that exhibit topographic scarps that may be the expression of recent surface deformation (Illies and Greiner, 1976; Cushing et al., 2000; Brüstle, 2002). Since a Pleistocene fluvial terrace of the River Rhine coincides with the location of the Rhine River scarp, Brüstle (2002) suggests that this scarp has a combined tectonic and an erosional origin.

In order to provide better constraint on the neotectonic activity and paleoseismicity of the northern URG, the geological and geomorphological records of fault movements were studied by integrating techniques in paleoseismology, structural analysis and shallow geophysics. Deformation observed in trenches was used to investigate evidence of surface deformation in young sediments. This study concentrated on the western part of the northern URG. Both border faults in the northern URG are nearly free of seismic activity; an exception is the Worms 1952 earthquake associated most likely with the WBF. The choice to focus on the WBF for paleoseismological investigations was mainly based on logistical conditions. In addition, it was expected to have Quaternary deposits, suitable for dating techniques, preserved on the hangingwall and footwall part of the WBF along its northern segment (adjacent to the Mainz Basin). In contrast, the footwall rocks of the EBF (basement in the Odenwald and Tertiary in the Kraichgau) were not suitable for dating recent tectonic deformation. The WBF segment chosen for paleoseismological trenching is part of a 20 km long, 50 – 100 m high topographic scarp (Fig. 2.2). The remarkably linear shape of this scarp resembles other fault scarps identified in the URG. The scarp has been highly modified by human activity, especially viniculture, dating back at least to Roman times. At several accessible sites along the scarp, pre-trenching surveys were undertaken. Based on these surveys a trench site at the southern end of the scarp was selected (Fig. 2.2, 2.3). This chapter presents the results of 4 trenches and high-resolution geophysical measurements (ground-penetrating radar, geoelectrical tomography, reflection seismics) and discusses their implications for recent URG tectonics.

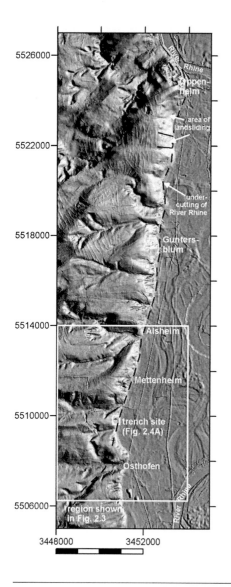

Figure 2.2. Shaded relief map of the 20 km long scarp investigated. The surface trace of the WBF (dashed line) is mapped by morphology (Steuer, 1911; Franke, 2001). South of Oppenheim two ~ E-W striking faults subsidiary to the WBF (solid lines; after Steuer, 1911) bound a landslide. Lateral resolution of the DEM is 20 m and vertical resolution is 0.5 m. Coordinates are in Gauss-Krüger coordinate system. Data from DGM5 (2002).

2.2. Geology of the Mainz Basin and the northern Upper Rhine Graben

The trenching site is situated along the contact of the western graben border with the Mainz Basin. In the following section, a short review of the Mainz Basin and northern URG, with emphasis given to the WBF and the investigated part of the 20 km long WBF segment, is presented.

2.2.1. The Mainz Basin

The Mainz Basin is bounded by three structural blocks (Fig. 2.1D): (1) In the North, the Mainz Basin is separated from the Rhenish Massif by the ENE-trending Hunsrück-Taunus Border fault, (2) to the southwest of the Mainz Basin, separated by a NW-SE trending fault contact, is the Nordpfälzer Hügelland and (3) in the east, bounded by the NNE-SSW trending WBF, the Mainz Basin borders the URG. From Late Eocene to the Early Oligocene the Mainz Basin and the URG developed as a single sedimentary system. Since the Late Oligocene, the Mainz Basin has subsided less than the URG and tilted southeastward in response to the episodic uplift of the Rhenish Massif (Meyer et al., 1983). Remnants of the Tortonian Dinotherium sands in the Mainz Basin are assigned to a paleo River Rhine (Bartz, 1936; Weiler, 1952; Grimm, 2002). Its course was across the Mainz Basin from Worms to Bingen (Bartz, 1936; Fig. 2.1D). Between the Early Pliocene and Early Pleistocene, relative uplift of the Mainz Basin forced the River Rhine to migrate northeastward to its present course along the Mainz Bingen Graben (Wagner, 1930; Kandler, 1970; Abele, 1977). During the Quaternary solifluction, fluvial erosion and loess deposition influenced the landscape evolution in the Mainz Basin. At present, the Mainz Basin has a tableland landscape formed by resistant Miocene limestones overlying Oligocene sands and marls. At present, the landscape is covered by Pliocene to Pleistocene fluvial sands and Upper Pleistocene loess and slope wash deposits.

2.2.2. The northern Upper Rhine Graben

The Cenozoic graben fill in the northern URG consists of a sequence of fluvial and lacustrine sediments, interrupted by marine sediments. The total thickness varies in this part of the graben from 400 m at Darmstadt to 3,200 m near Worms (Doebl and Olbrecht, 1974). The fluvio-lacustrine Quaternary deposits of the northern URG show large thickness variations from east (~ 380 m in the *Heidelberger Loch*) to west (40 m along the WBF segment investigated; Straub, 1962; Doebl, 1967; Bartz, 1974; Haimberger et al., 2005), reflecting the asymmetry of the graben structure (Fig. 2.1C). This thickness variation is the result of continuous subsidence of the eastern part of the URG, localized in the *Heidelberger Loch*. In contrast, the relatively low subsidence of the western side of the graben is accommodated by several fault branches of the WBF, leading to a more widespread distribution of deformation.

Several fluvial terraces of the River Rhine system developed in the northern URG during the Quaternary. The distribution of these terraces is different in distinct areas. Only small remnants of Early Pleistocene terraces are preserved on the eastern side. In contrast to this, widespread Early to Late Pleistocene terraces exist on the western side in the Vorderpfalz at the foot of the Pfälzer Wald (Stäblein, 1968; Monninger, 1985), and a complete sequence of Pleistocene terraces is present on both sides of the Mainz Bingen Graben (Kandler, 1970; Abele, 1977). The Lower Terrace, which was formed during the last glacial period (Würm), extends from the eastern to western border of the northern URG. During the Holocene, the River Rhine incised into the Lower Terrace (Fetzer et al., 1995; Dambeck and Thiemeyer, 2002).

2.2.3. The Western Border Fault (WBF)

The WBF strikes in a NNE-direction from Hofheim a.T. at the northern end of the graben southward across Rüsselsheim, Oppenheim, Osthofen towards Bad Dürkheim (Fig. 2.1D). Along this section, it separates the Mainz Basin from the URG. Several sub-parallel faults reflect the complex deformation history of the western margin of the URG. The vertical offset along the WBF is highly variable (Fig. 2.1D). In the central part of the scarp segment investigated (Mettenheim), Lower Miocene Hydrobia Beds are offset by 824 m (Franke, 2001). Southwards, the vertical offset of Hydrobia Beds on the WBF gradually decreases to 122 m with the extension taken up on the Worms Fault, which accommodates offsets of Hydrobia Beds to a maximum of 1,035 m (Stahmer, 1979). The vertical movements on both faults occurred mainly during the Miocene, contemporaneous with the major depositional phase (2,000 m of the maximum 3,200 m sediment fill was deposited during the Miocene; Doebl and Olbrecht, 1974). Along the eastern border fault Quaternary tectonic activity remained similar to its Late Tertiary level. The relatively thin accumulation of Quaternary sediments (40 – 100 m; Fig. 2.1C) along the WBF suggests minor activity along this structure. In contrast, the amount of Quaternary uplift of the adjacent Mainz Basin (170 m; Brüning, 1977) and the formation of fluvial terraces on the western margin imply a significant vertical displacement along the WBF.

2.2.4. The WBF scarp

The 20 km long scarp along the WBF segment investigated (Fig. 2.2, 2.3) is a striking morphological feature. It has a height of 100 m in the north and 50 m in the south relative to the present River Rhine flood plain. The time of formation and origin of this scarp is unclear and could be fluvial, tectonic, or a combination of both. Since the Early Pleistocene, the River Rhine migrated from the uplifting Mainz Basin to the Mainz Bingen Graben (Wagner, 1930; Kandler, 1970; Abele, 1977; Fig. 2.1D). It is likely that scarp formation was initiated at this time but there is no evidence to prove this hypothesis.

Based on enhanced seismic activity and precise leveling data, Illies (1974a) suggests that the WBF between Hofheim and Oppenheim, north of the segment investigated, is an active fault zone (Fig. 2.1B, D). In contrast, precise leveling data from Schwarz (1976) show enhanced relative vertical movements occurring along the WBF segment investigated. These measurements indicate vertical movement rates increasing from 0.24 mm/yr in the northern part to 1.26 mm/yr in the southern part.

In the northern part (Oppenheim to Alsheim), the scarp exhibits several indentations caused by land sliding and fluvial erosion of the River Rhine (Fig. 2.2). At Oppenheim, landslides have been frequent in prehistorical and historical times (e.g. Steuer, 1911; Wagner, 1941a; Krauter and Steingötter, 1983). These landslides are related to the lithological conditions at this part of the scarp, where the Lower Oligocene sands tend to flow once they become water saturated. Here, the largest landslide is located between two ~E-W trending faults (Steuer, 1911) indicating that fault structure has some influence on landform evolution. During the Middle Würm, the River Rhine eroded into the scarp near Guntersblum and triggered a riverbank collapse (Semmel, 1986; Fig. 2.2). In recent times, human activity, especially for viniculture, has caused instabilities of the slope in this part of the scarp.

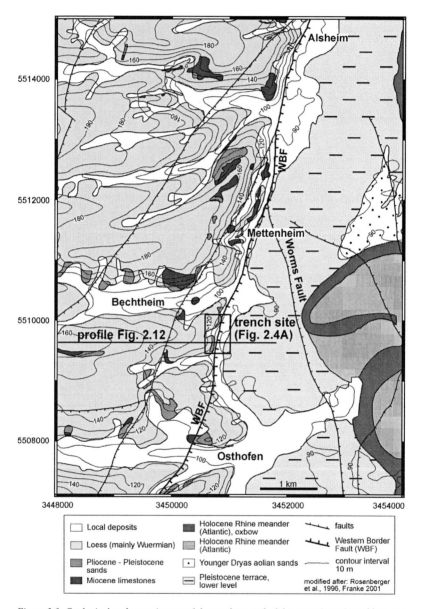

Figure 2.3. Geological and tectonic map of the southern end of the scarp investigated between Alsheim and Osthofen. At the trench site shallow geophysical measurements and corings were

undertaken. The pre-survey results indicated this site as the most favorable for trenching. Additionally, the location of a west to east topographic profile given as Figure 2.12 is shown.

The southern end of the scarp (Alsheim to Osthofen) is the most linear (Figs. 2.2, 2.3). Here, landslides are less likely since the scarp is lower and less steep. Furthermore, the scarp is mainly formed by resistant Miocene limestones covered with Pliocene – Pleistocene sands, alluvial loess and slope wash (Fig. 2.3). Land consolidation has been less intense than in the north and no recent mass movements have been reported. In addition, this part of the scarp has the most natural morphology. For these reasons it was thought to be the best location for the preservation of near-surface deformation along the WBF.

2.3. Pre-trenching surveys

Before focusing with pre-trenching surveys on the southern part of the WBF scarp, six other locations in the northern URG have been investigated for potential trenching locations. A summary of all pre-trenching surveys undertaken in the framework of this thesis is given in the Appendix (see Figure A2.1 for locations of survey sites). Parts of the survey and trenching results have also been documented in Wenzel et al. (2004). In the Appendix, the sites and the measurements undertaken on both sides of the URG are presented and the reasons for not choosing these sites are discussed. The review of the geology of the northern URG (section 2.2) demonstrated the asymmetric graben morphology and showed that the largest vertical displacements along the border faults occur at Heidelberg along the EBF. Therefore, field reconnaissance concentrated first on the EBF. However, since the entire eastern border from Karlsruhe in the south to Darmstadt in the north was highly urbanized and the EBF scarp was nearly nowhere accessible, this part of the graben was not ideal for geophysical investigations and trenching. For this reason, subsequent field surveys concentrated on the less populated western side of the northern URG. In addition, it was expected to have Quaternary deposits, suitable for dating techniques, preserved rather along the WBF than the EBF.

For trenching, the trace of the fault at surface needs to be known to within a few meters. In the case of the WBF segment, its location is only known at depth from industry seismics and cores. The faults' surface trace has been mapped assuming that it coincides with the base of the scarp (Steuer, 1911; Franke, 2001; Fig. 2.2, 2.3). High-resolution geophysical imaging techniques (reflection seismics, geoelectrical tomography and ground-penetrating radar) were used to map the fault at surface accurately. In addition, shallow coring was performed at several locations. The site chosen for intense geophysical measurements and trenching was situated north of Osthofen and covered an area from the base of the scarp to the first vineyard terrace 10 m above (anthropogenic V1, Fig. 2.3, 2.4). The base of the scarp corresponds to the Lower Terrace of the Würm glacial period (Franke, 2001; Fig. 2.4A). The presence of older fluvial terraces on the scarp at this location has not been previously investigated. It is therefore possible that V1 also corresponds to a fluvial terrace. Geophysical measurements and drilling were undertaken on fallow vineyards and paths. The groundwater table at the trench site was reached with coring and located at 6 – 8 m below the surface at the base of the scarp (campaigns in spring to fall of 2002 and 2003). On V1, the groundwater was too deep to be reached with coring. Due to the relatively dry seasons of the years 2002 and 2003, the groundwater was not detected with GPR or resistivity measurements. In the following section and on Figure 2.5, the results of the geophysical measurements and corings are presented and the many indications for near-surface structures that the surveys provided are discussed.

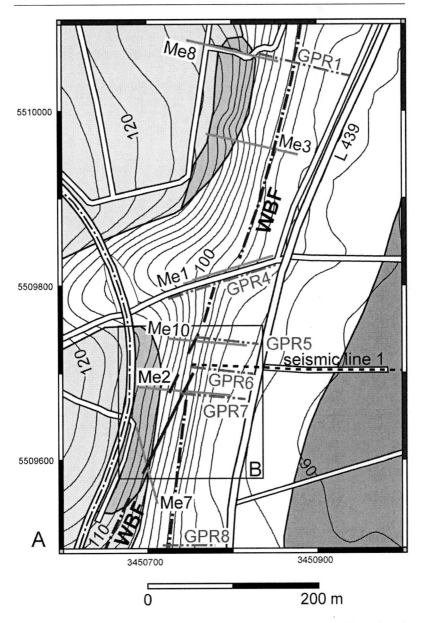

Figure 2.4. A) Geological map of the trench site chosen, with location of all geophysical profiles performed and inferred surface trace of the WBF. Grey solid lines: geoelectrical profiles Me1, 2, 3, 7, 8, 10. Geology after Franke(2001). Stippled dark grey lines: GPR profiles

GPR 4 – 8. Black dashed line: seismic line 1. Solid and dashed black line: WBF surface trace, known and inferred (legend as Figure 2.3).

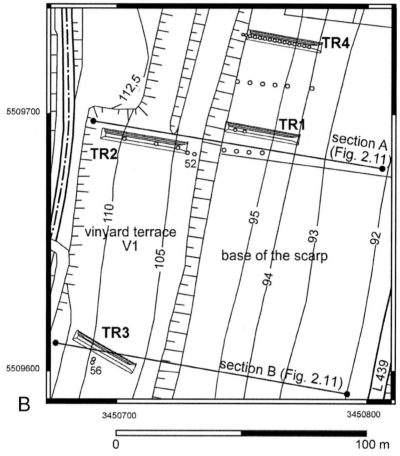

Figure 2.4. B) Detailed topographic map of the trench site showing the location of cores (open circles), trenches and restored sections given as Figure 2.11. Topographic map after DK5 (1980).

2.3.1. Description of the geophysical profiles

2.3.1.1. High-resolution reflection seismics

A drop weight of 50 kg and a geophone spacing of 1 – 2 m were used for a seismic profile with a penetration depth of about 200 m (seismic line 1 on Figures 2.4A, 2.5A). Unfortunately, measurements on the steep slope of the scarp were not possible. The processed seismic section is shown in Figure 2.5A. In the eastern part of the section, reflectors are continuous and horizontal (meter 240 – 400). In the middle and western part of the section (meter 100 – 240), reflectors are interrupted at four zones, which are interpreted to be high angle extensional faults (F1 – F4). Fault F3 extends almost to the surface, whereas the other three faults probably terminate ~ 40 – 50 m (0.025 – 0.06 s TWT) below the surface. The uppermost reflectors in the western part form a wedge shaped structure (meter 100 – 200) indicating westward thickening of the strata towards the graben border fault. This structure is interpreted as a growth sequence associated with a half graben, of which the bounding, east dipping, synthetic fault (F5; Fig. 2.5A) is thought to be located to the west of the profile. This interpretation is supported by the increase in fault structures and decrease in spacing between the faults in the western part of the section.

2.3.1.2. Geoelectrical tomography profiles

Several 2D-geoelectrical tomography profiles using a Wenner array have been carried out. The profiles have a resolution of 2 x 2 m and reach depths of 20 – 30 m. The resistivity of the ground depends on several geological parameters, including porosity, mineral content and fluid saturation (e.g. Friedel, 1997; Loke, 2001).

Five geoelectrical profiles (Me1, 2, 3, 8, 10) cross the scarp and one profile (Me7) is located on V1 (Fig. 2.4A). Low resistivity (< 145 Ωm) suggests that only soft sediments occur in the upper 30 m (Fig. 2.5B). On all profiles a distinct inclined boundary between relatively high and low resistivity occurs (Fig. 2.5B). The surface projection of this distinct change in resistivity is mapped in Fig. 2.4A. In the north, this zone is parallel to the scarp whereas in the south it crosses the scarp.

Next page:

Figure 2.5. A) Processed and interpreted results of seismic reflection line 1. B) Geoelectrical profiles showing resistivity contrasts along a subvertical, east dipping surface. C) GPR profiles with disturbed and steeply dipping reflectors in a 30 m wide zone eastwards of the base of the scarp. Dashed line: maximum penetration depth. Box in GPR 7 indicates the location of TR1.

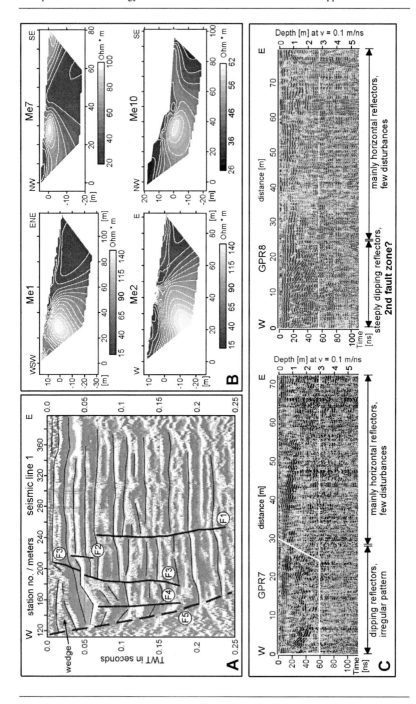

2.3.1.3. Ground-penetrating radar (GPR)

The reflection pattern of high frequency electromagnetic waves detected by GPR can represent lithological boundaries, moisture content or fault structures in the shallow subsurface. A 200 MHz antenna was used to investigate up to 3 m deep at 0.2 m resolution. A series of parallel profiles crossing the base of the scarp were made (Fig. 2.4A). Six of these profiles exhibit similar reflection patterns: an approximately 30 m wide zone extends eastwards from the base of the scarp in which reflectors are steeply eastward dipping and frequently disturbed (Fig. 2.5C). The eastern parts of the profiles show undisturbed, subhorizontal reflectors.

2.3.1.4. Percussion cores

In total 34 percussion cores (63 mm diameter) were taken in order to obtain a three dimensional picture of the various lithological units at the trench site. These cores were located both on V1 and at the base of the scarp where they reach a maximum depth of 8 m (Fig. 2.4B). Cores obtained at the base of the scarp in general reveal a uniform stratigraphy. Fine sand and silt deposits with a thickness of 4 – 5 m are overlying a sequence of cohesionless sands and clay-rich sands that continue to the maximum coring depth. No clear evidence for faults could be derived from the cores at the base of the scarp because of the heterogeneity of the lower sequence with the various layers thinning out and/or forming lenses. However, it is interesting to note that the largest variations in thickness of the sand layers occur near the base of the scarp, i.e. where the geoelectrical and GPR measurements indicate resistivity contrasts and deformed structures respectively.

In contrast to the sequence found at the base of scarp, cores taken on V1 show a different stratigraphy. On V1, a thin unit of cohesionless sand overlies 11 m thick clay-rich sand, which forms this part of the scarp. This unit overlies the sequence of cohesionless sands and clay-rich sands found in cores at the base of the scarp.

2.3.2. Trenching locations

Integration of all geophysical measurements and core data suggest a zone of sediment deformation located along the strike of the scarp (Figs. 2.4A, 2.5). The origin of these deformations (i.e. tectonic or sedimentary) cannot unambiguously be derived from the surveys. For this reason, trenched profiles were made in order to obtain further data on these structures. Trench locations were selected on the base of the results of these surveys and site conditions.

The first trench, TR1, was located at the base of the scarp (Fig. 2.4B) along the two most promising geophysical profiles (Me2 and GPR7, Fig. 2.5B, C) and where thickness variations in cores were largest. Here, a fault zone was discovered. A second 30 m long trench, TR2, was situated on V1 in the westward continuation of TR1, leaving a gap between the trenches (Fig. 2.4B). This trench was opened in order to gain information on the sedimentary evolution of the scarp and on the westward extension of the fault zone uncovered in TR1. Since geophysical profiles and the structural analysis of the fault zone indicated that one fault splay crossed the scarp, a third 20 m long trench, TR3, was opened ~ 100 m south of TR2 on V1 (Fig. 2.4B). In order to follow the trace of the fault zone to the north of TR1 a fourth trench, TR4, was opened ~ 30 m north of TR1.

2.4.   Trench descriptions

The stratigraphy of the trenching area has been determined by detailed mapping of the trench walls (1:10 mapping), percussion coring, absolute dating and lithostratigraphic correlation. In total 10 stratigraphic units have been identified and are presented in detail in the Appendix of this thesis (section A.3) and Table 2.1. This section describes the individual trenches. It is followed by a structural analysis and interpretation of the faults uncovered in the trenches. Finally, a reconstruction of the site is presented.

2.4.1. Trench 1

2.4.1.1. Overview

TR1 is located at the base of the scarp (Fig. 2.4B) along parts of the geophysical profiles Me2 and GPR7 (Fig. 2.5B, C) and where thickness variations in cores were largest. TR1 exhibits deformation caused by faulting (Figs. 2.6D, 2.6E), liquefaction features (Fig. 2.7) and soft sediment deformation related to mass movements (Fig. 2.6A, inset IV). Most of the tectonic deformation is located in a 6 m wide fault zone in the western part of the trench.

TR1 exhibits 6 of the stratigraphic units identified (Fig. 2.6). Unit 1A (sand with clay lenses) occurs in the westernmost basal part of the trench. Above unit 1A lays unit

3A, a conglomerate, and unit 3B of interbedded sand and silts. Both unit 3A and 3B dip to the east. Unit 4A (iron coated sand and silt) has a steeply, eastward dipping erosional discontinuity on the lower units (1A, 3A, 3B). In the middle part of the trench, unit 4A is horizontally bedded and lies on unit 1A. Unit 4B (alluvial loess) conformably overlays unit 4A. Recent colluvium (unit 5A) and a thin Holocene horizon (unit 5B) truncate units 3B and 4B.

Unit		Interpretation	Lithology	Sedimentary structures	Base	Remarks	Color (Munsell code)
5	B	soil	silts, sands, organic material	bioturbation		carbonate-rich, average slope of 8° in TR1, 14° in TR2 and TR3	greyish brown 10YR 5/2, dark greyish brown 2.5Y 4/2
	A	colluvium	sands, silts, gravels, carbonate concretions		erosive	partly carbonate-rich, detrital charcoal fragments, slag, plastic, weakly developed soils, Middle Age and Roman artifacts, slope 12 – 19°	dark yellowish brown 10YR 4/4, yellowish brown 10YR 5/4, brownish yellow 10YR 6/6
4	B	alluvial loess, slope wash deposit	silts, fine sands, partly medium sands and course sands	parallel lamination (mm-cm scale)	erosive ?	quartz and mica-rich, mass movement features, liquefaction features, slope 11 – 28°	pale yellow 2.5Y 7/4
	A	slope wash deposit	medium sands, silt	lamination (cm scale)	erosive	iron coating on sand grains, slope 5 – 24°	yellowish brown 10YR 5/8
3	B	slope wash deposit	fine sands, silt	parallel lamination (mm-cm scale), sandy fan deposits with fining upward	sharp and erosive	quartz and mica-rich, rounded limestone pebbles and one flattened (20cm x 5cm) silt clast at base, average slope 20°	light yellowish brown 2.5Y 6/4
3	A	basal conglo-merate	gravel, limestone clasts in sandy matrix (upper part) and clayey matrix (lower part), clay layers	lenticular bedding (cm scale)	erosive	Miocene limestone fragments with sharp edges, rounded quartz and sandstone pebbles, manganese and iron nodules, carbonate concretions, slope 9° – 12°	strong brown 7.5YR 5/8
2	B	Paleocambisol	sands in clayey matrix	bioturbation	gradual	manganese nodules and layer at the base, no slope	yellowish brown 10YR 5/6
	A	braided river deposit, terrace body	fine-medium sands, course sand, gravel, gravel layer at the base	crossbedding and parallel lamination (cm scale)	erosive	cohesionless, carbonate-free, quartz pebbles, 10 cm thick gravel layer with pebbles, hematite nodules, wood pieces at base, Buntsandstein provenance (Pfälzer Wald), no slope	pale yellow 2.5Y 8/2 brownish yellow sandlayers: 10YR 6/8
1	B	fluvial or lacustrine still water deposit	clayey, fine-medium sands		sharp	pseudogley formation, manganese layers and nodules, mottled, Buntsandstein provenance (Pfälzer Wald) with zirkon, turmalin, rutil, anatas, no slope	yellowish brown parts : 10YR 5/8; light grey parts: 2.5Y 7/1
	A	braided river deposit	fine-medium sands, clayey sands, clays	lenticular bedding (dm-m scale)	?	carbonate concretions, upper part carbonate-free, lower part carbonate-rich, rounded quartz grains, liquefaction features, slope 1 – 8°	pale yellow sands: 2.5Y 8/2; light grey clays: 2.5Y 7/2

Table 2.1. Part1: Description of lithostratigraphic units identified in the trenches and interpretation of their formation environment.

Unit		Period	Estimated age (ka)	TL and C14 dating (ka), heavy mineral analysis samples
5	B	Holocene, MIS 1	10 – recent	MET14: 5.83+/-0.76 MET33: 10.64+/-0.95 MET36: 8.48+/-0.93
	A	Late Holocene (since Middle Age – recent)	~0.75 – recent	MET54: 3.43+/-0.28 MET56: 2.56+/-0.32 MET57: 2.13+/-0.23 MET58: 0.41+/-0.06 MET115: 24.3+/-2.5 MET118 : 3.98+/-0.51 MET131: 23.3+/-3.8 Metten-Tr.1-2-SW/C14: 44.46+1.17/-1.02 BP Metten-Tr.1-6-SW/C14: 2.64+/-0.025 BP Metten-Tr.1-11-SW/C14 : 0.207+/-0.021 BP
4	B	Late Würm	23 – 14	MET32: 18.9+/-2.4 MET34: 14.6+/-1.5 MET35: 14.2+/-1.3 MET52: 19.4+/-2.0 MET53:14.2+/-1.5 MET55:14.5+/-1.9 MET119: 23.4+/-2.5 MET120: 22.7+/-2.0 MET133: 18.2+/-1.2 MET134: 17.9+/-1.4 MET135: 19.4+/-1.4 MET136: 19.9+/-1.8
	A	Middle – Late Würm	31 – 22	MET31: 30.8+4.1/-3.6 MET81: 22.2+/-2.7
3	B	Early – Middle Würm?, MIS 5b – 3	?100 – ~37	MET12: >100 MET13: 19.4+/-1.9 MET132: 37.3+4.6/-3.9
3	A	Early Würm?, MIS 5b	?100	
2	B	intra & post ? Eemian, MIS 5e	?130 – ?recent	MET111: >100 MET112: >100 MET114: >100
	A	Riss terrace?, MIS 6	?450 – ?130	MET113: >100 heavy mineral analysis, sample no.: MET-T4-01; MET-T4-02; MET-T4-03; MET-T3-03; MET-T3-02; MET-T3-01
1	B	Pliocene?	?>3000	heavy mineral analysis, sample no. and depth: MET-BK56-03: 3.60-3.93 m; MET-BK56-05: 5.50-5.92 m; MET-BK56-10: 7-8 m; MET-BK56-11: 8-9 m, depths counting from top of unit 1A
	A	Pliocene?	?>3000	MET11: >100 MET51: >100

Table 2.1. Part 2: Estimated and absolute ages of lithostratigraphic units.

2.4.1.2. Faulting

The fault zone consists of a large number of mainly extensional faults with offsets of up to 0.60 m (Figs. 2.6D, 2.6E). A minor fault set of strike-slip faults with offsets of a few centimeters has been observed on the west wall (Fig. 2.6C). On the footwall side of the fault zone, both synthetic and antithetic faults occur. The synthetic faults dip steeply (~ 60°) to the east. The antithetic faults are generally shorter, especially on the north wall. They dip steeply to the west (~ 65°) and abut the synthetic faults. The core of the fault zone is defined as the synthetic fault with the largest vertical throw (0.6 m) and is termed the main fault (thick solid line on Figures 2.6D, 2.6E). The hangingwall consists of a 1 m wide zone of mainly synthetic faults. Note that the hangingwall and footwall exhibit a different fault style. Furthermore, the fault core is located in different stratigraphical positions on the north and south walls. On the south wall, it is located mainly at the erosional contact between unit 3B with 4A (Fig. 2.6D) suggesting that faulting has used the erosional contact. On the north wall the fault core is entirely situated within unit 3B (Fig. 2.6E). The fault density is related to lithology. A high fault density exists in the sandy and silty units (1A, 3B, 4A, 4B) while a lower density occurs in the clay-rich conglomerate (3A). On the south wall, a horst structure is situated above the lower, inclined part of the reactivated erosional surface that indicates slip on an irregular surface (Fig. 2.6D).

Next page:

Figure 2.6. A) TR1 trench log (based on 1:10 mapping) of the south wall showing the lithostratigraphic units, faults, decollements and absolute ages from dating.

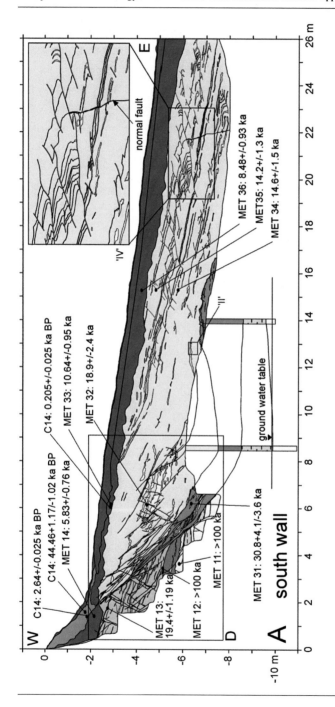

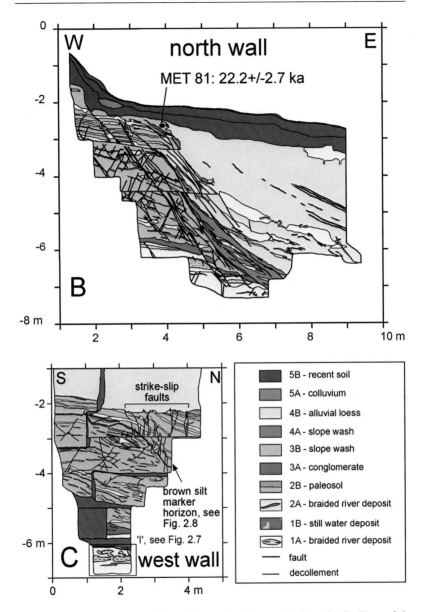

Figure 2.6. B) and C) TR1 trench log of the north and west walls. Legend is for Figures 2.6, 2.8, 2.9 and 2.10.

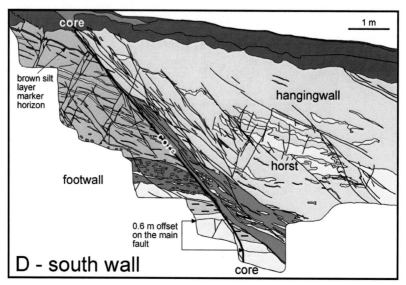

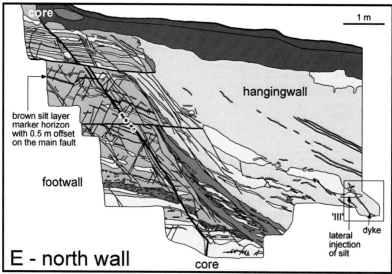

Figure 2.6. D) and E) Insets D and E show the fault zone in detail and define the footwall, main fault (fault core) and hangingwall.

2.4.1.3. Liquefaction features

Liquefaction features, indicating injection of water-saturated material under overpressure, are observed at several locations in TR1. They are restricted to the sandy and silty sediments of units 1A and 4B. On the base of the west wall, cohesionless sand has been injected into overlying clay-rich sand, both of unit 1A, vertically and laterally (Fig. 2.6C, marked 'I', Fig. 2.7). A 5 cm thick clay layer overlying these structures acted as a seal. Upwardly injected silt occurs as bubbles within sand-rich silt (Fig. 2.6A, marked 'II'). On the north wall, a 1 m long silt dyke exists (Fig. 2.6E, marked 'III'). Wavy bedding, micro-flame structures and stretched sand layers in the alluvial loess are frequent on both the north and south walls. Such features and the combination of their occurrence have been described as liquefaction features induced by seismic events (e.g. Obermeier, 1996a; Obermeier, 1996b, see Figure 7.23 in here; Tuttle, 2001).

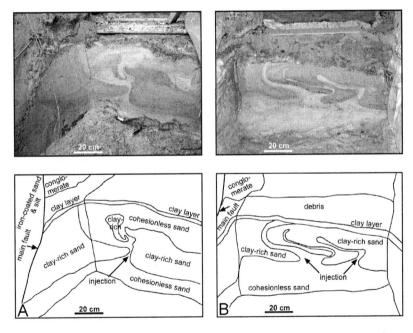

Figure 2.7. Photos of liquefaction features at the base of TR1. Photo B was taken at a later stage of excavation, when the wall was located approximately 20 cm further west than at the time of photo A.

2.4.1.4. Mass movement features

From the central to the eastern part of TR1 (Fig. 2.6A meter 12 – 26, inset marked 'IV'), decollements with a slight westward dip occur in the alluvial loess. These decollements are rooted in a fine sand layer and sand has been injected along them. Bedding is crosscut by these structures, indicating they formed after sediment deposition (i.e. post 14 ka; MET34; Table 2.1, Fig. 2.6A). The offset along these structures is extensional in the lower part and compressional (east vergent folding) in the upper part. All these features strongly indicate a mass movement. However, the origin of this mass movement is ambiguous. The alternating process of thawing and freezing of the top of the water-saturated soil located above impermeable permafrost during the Late Würm could have created conditions favorable for gravitational movements on the existing slope. It is also likely that fault activity may have triggered the mass movement observed. Extensional movements on the nearby faults could have possibly affected the stability of the slope.

2.4.2. Trench 2

2.4.2.1. Overview

TR2 is situated west of TR1. It was made in order to investigate the westward extension of the fault zone discovered in TR1 and to gain information on the sedimentary evolution of the scarp. TR2 exhibits only one single extensional fault structure in the eastern part of the trench (Fig. 2.8A). In this part, the westward continuation of the units 3B and 4B from TR1 is exposed (Figs. 2.6A, 2.8A). These sloping units overlie the 11 m thick unit 1B (clay-rich sand), which forms this part of the scarp.

2.4.2.2. Faulting

A steeply eastward dipping (62°) extensional structure, consisting of a few fault strands, is situated within units 1B, 3B and 4B (Fig. 2.8A, inset marked 'I'). All fault strands terminate within unit 4B (alluvial loess). Adjacent to the faults, a few folds and decollements, showing the same deformation style as in TR1, are seen (Fig. 2.8A, inset marked 'I').

2.4.2.3. Sedimentary structures

Units 1B and 2A were deposited on a subhorizontal surface and at a later stage eroded from the eastern part of TR2. On the erosional and eastward sloping surface of unit 1B units 3B and 4B were deposited (Fig. 2.8A). During historical times the slope was modified, most likely for a vineyard terrace, and was covered by an almost 2 m thick colluvium (unit 5A).

2.4.3. Trench 3

2.4.3.1. Overview

TR3 is also situated on V1 about 100 m south of TR2. The stratigraphic position of TR2 and TR3 is therefore very similar (Fig. 2.8). The results of the structural analysis of the fault zone discovered in the first trench indicated that one part of the fault zone crossed the scarp and extended on V1 in a NNE-SSW orientation (Fig. 2.4A). The geoelectrical profile Me7 showed a distinct resistivity contrast at the position of the extrapolated fault zone (Figs. 2.4A, 2.5B). Based on this information, TR3 was opened in order to verify the location of a fault zone. The trench log shows that TR3 exposes a 5 m wide fault zone in the central part of the trench (Figs. 2.8B, 2.8C).

2.4.3.2. Faulting

The fault zone deforms the units 1B, 2A and 2B (Figs. 2.8B, 2.8C). Similar to the fault zone in TR1, this zone consists of numerous steeply dipping, extensional synthetic and antithetic fault strands. In the west, the footwall exhibits multiple large synthetic faults (Fig. 2.8C). Fewer shorter antithetic fault strands abut the synthetic faults. The fault core is situated along a synthetic fault structure consisting of multiple, closely spaced fault strands (thick solid lines on Figure 2.8C). The strands merge into a single fault towards the top of the fault zone where a remnant of a paleosol, unit 2B, is faulted against the sands of unit 2A. The hangingwall deformation consists mainly of antithetic faults and a few synthetic faults. Here, the antithetic faults are less steep than in the footwall. Similarly to TR1, fault density is clearly related to the lithology. There is a high fault density in the sandy unit 2A, whereas a lower density occurs in the clay-rich unit 1B. In order to investigate the faulted surface in 3 dimensions the base of the trench was carefully excavated to expose the top surface of unit 1B. In addition to the extensional faults known from the trench walls, a minor set of sinistral strike-slip faults that offset the top surface by a few centimeters was exhumed during excavation.

Next page: Figure 2.8. A) Trench log of TR2.

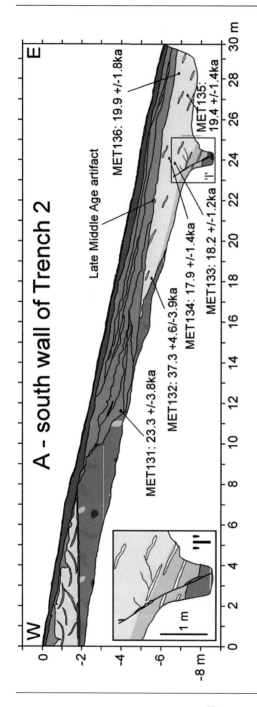

A - south wall of Trench 2

MET136: 19.9 +/-1.8ka

MET135: 19.4 +/-1.4ka

Late Middle Age artifact

MET133: 18.2 +/-1.2ka

MET134: 17.9 +/-1.4ka

MET132: 37.3 +4.6/-3.9ka

MET131: 23.3 +/-3.8ka

1 m

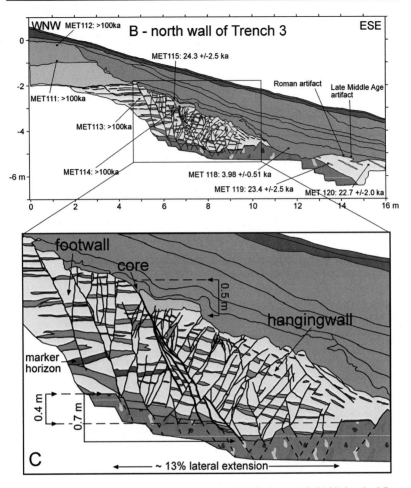

Figure 2.8. B) and C) Trench log of TR3. The inset of the fault zone (C) highlights the 0.7 m vertical displacements and the net 0.4 m displacement of the top of unit 1B. Unit 2B is offset a minimum of 0.5 m. Due to lateral variations, the marker horizon (red) cannot be correlated across the main fault and its' fault strands (fault core, thick black lines).

2.4.3.3. Sedimentary structures

The sedimentary pattern in TR3 is very similar to TR2. Units 1B and 2A are subhorizontally bedded while unit 4B deposited on a sloping surface. The thickness (maximum 2.5 m) of unit 5A, a colluvium of historical age, suggests active man-made modifications at this site.

2.4.4. Trench 4

2.4.4.1. Overview

TR4 is located ~ 30 m north of TR1. This trench was opened in order to follow the trace of the fault zone to the north of TR1. TR4 was however a failure in as much as it revealed only small extensional faulting. The stratigraphy of TR4 is for the lower part of the trench similar to TR1 (Fig. 2.9). Cores taken from the bottom of TR4 reveal unit 1A. This is overlain by unit 4B. Most of TR4 exposes a thick sequence of soils and colluvium (unit 5A). The lower part of this sequence is a humus rich soil that replaced unit 4B. After R. Dambeck and M. Weidenfeller (pers. comm., 2003) this soil is a chernozem (black soil) of possibly Boreal age (~ 8 ka).

2.4.4.2. Faulting

Small scale extensional faulting occurs in units 1A and the lower part of 4B in the western and eastern parts of the trench (Fig. 2.9). Vertical offsets in unit 1A reach 10 cm while they are reduced to 4 cm in unit 4B. The younger soil horizon is undeformed. The concentration of fault structures in the western part of the trench suggests the presence of a larger fault zone nearby. This is supported by the geoelectrical profile taken at the position of the trench and further to the west, that shows a distinct resistivity contrast west of the trench extensions (Fig. 2.5B, Me10).

2.4.4.3. Sedimentary structures

Interpretation of several cores reveals distinct lenticular bedding of unit 1A (Fig. 2.9). Like in the other three trenches unit 4B deposited also at this site on a sloping surface and shows mildly east dipping bedding. The thick colluvium on top of the chernozem horizon indicates active man-made modifications.

Next page:

Figure 2.9. Trench log of TR4.

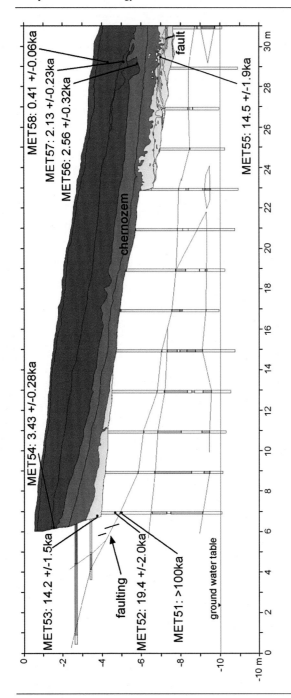

2.5.   Structural analysis and timing of the faulting

2.5.1. Extensional displacements

The majority of extensional, vertical displacements on single fault strands in all four trenches are in the order of a few centimeters. A few faults (main faults or fault core) show vertical offsets of ~ 0.5 m. In TR1, these large offsets are on the south wall within unit 1A and on the north wall within unit 3B (Figs. 2.6D, 2.6E). Due to poor correlation of units across the erosional discontinuity, which reactivated as a fault surface, these offsets are the maximum offsets that can reliably be determined. In TR2, the maximum vertical offset is only 0.17 m and in TR3, the maximum vertical offset is the accumulated offset of several synthetic faults of 0.7 m (Figs. 2.8A, marked 'I', 2.8C). Due to large offsets on antithetic faults in TR3 a graben structure is developed. The net vertical offset of this graben structure, determined by line length restoration of units 1B and 2A, is 0.4 m. In comparison, the fault zones in TR1 and TR3 show both a maximum vertical offset in the order of 0.5 m. None of the offsets in the trenches give evidence for synsedimentary faulting, which implies that the deformation occurred entirely post depositional.

2.5.2. Timing of the faulting event

The timing of the faulting event can only be constrained by determining a maximum and a minimum age of the affected sediments. The youngest datable sediments affected by faults are unit 4B in TR1, TR2 and TR4 (Figs. 2.6, 2.8, 2.9). Thermoluminescence (TL) dating of samples located close to the fault structures yields an age of 19 ka (MET32, median age of MET133, 134, 135, 136, MET52; Table 2.1). If the single extensional fault, located at the eastern end of TR1 (Fig. 2.6A, marked 'IV'), was active simultaneously with the main fault zone at the western end of the trench, a maximum age would be 14 ka (derived from sample MET35; Table 2.1). Unfortunately, simultaneous activity cannot be proven. The same accounts for the small faults in TR4, which are located near 14 ka old deposits. Since the dominant fault structures affect 19 ka old units, this more conservative estimate is chosen as the maximum age for the faulting. The minimum possible age is obtained from dates of the soil horizon (unit 5A) truncating the faults in TR1 and the chernozem in TR4 (Figs. 2.6, 2.9). TL dating of the soil in TR1 yields a median age (3 samples) of 8 ka (Table 2.1). This age is also suggested for the chernozem (R. Dambeck and M. Weidenfeller, pers.

comm., 2003).

2.5.3. Structural data

Structural analyses of the fault structures in all four trenches have been performed in order to compare the style of deformation between the different trenches and to relate the fault system at the trench site to active regional fault trends. If the fault structures in the trenches are part of the same fault system at a larger scale, a similar style of deformation with similar orientation of fault sets is to be expected.

The data of TR1 show a dominant set of conjugate, extensional faults with a NNE-SSW strike direction (maxima at 015°; Fig. 2.10A). Synthetic faults are more frequent than antithetic faults. Subsets are a set of NW-SE (120°) striking sinistral strike-slip faults (great circle 3) and an ENE-WSW (075°) striking, oblique to dextral strike-slip set (great circle 4). Measurements of the fault strands in TR2 yield a 005° striking set of synthetic faults and a few antithetic faults (from the north wall; Fig. 2.10B). The structural data from TR3 reveal a set of conjugate NNE-SSW (015°) striking extensional faults (Fig. 2.10C, great circle to 1 and 2). In the cohesive clay-rich sand (unit 1B), these faults show only minor variations in orientation. Conversely, in the cohesionless sand (unit 2A) the faults 'fan out' and have strike orientations between 015° and 065° (Fig. 2.10C, 'II'). Subsets are 065°-striking extensional faults (great circle 4) and 115°-striking, sinistral strike-slip faults (great circle 3). The latter set was identified on the surface of unit 1B. The small number of extensional faults of TR4 consists of two sets of NE-SW and NW-SE strike (Fig. 2.10D).

The structural analysis shows identical fault trends in all trenches. When plotted together (Fig. 2.10E, F) the data yield a dominant, conjugate fault set of NNE-SSW striking extensional faults (great circle 1 and 2) and three minor subsets (great circles 3, 4 and 5). The average fault dip is 55°. In addition to the similarities in fault trends, the maximum vertical offsets are consistent with ~ 0.5 m. These observations strongly suggest that the fault zones in TR1 and TR3 plus the small fault structures of TR2 and TR4 belong to the same fault system.

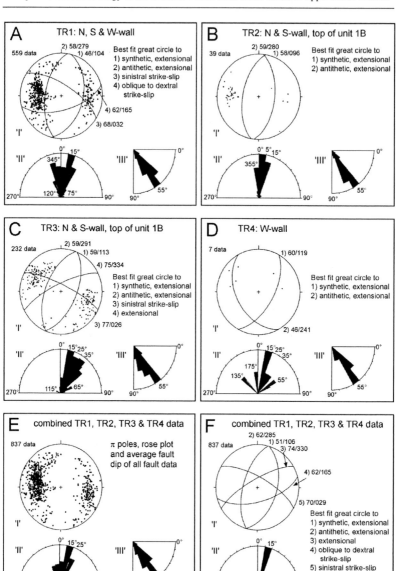

Figure 2.10. Fault data from TR1 (A), TR2 (B), TR3 (C) and TR4 (D) using data of all the trench walls and from the floors of TR2 and TR3: I) π poles to faults plotted on lower hemisphere, equal area projection and best fit great circles through dominant fault sets. II) Rose plot of fault data with 1° of maxima. III) Inclination plot of faults. E: π poles, rose plot

and average fault dip of all fault data. F: Summary plot of all data with sets of best-fit great circles. In order to ensure as high a quality data set as possible, only fault segments are included in the analysis on which two or more consistent measurements where recorded.

The NNE-SSW (015°) trend of the fault zones is the so-called Rhenish trend, which is represented by the border faults of the URG. Studies of outcrops and reflection seismic profiles give indications for two principle trends of neotectonic faults in the northern URG: (1) N-S trend (Straub, 1962; Anderle, 1974; Schneider and Schneider, 1975; Illies and Greiner, 1976), and (2) NE-SW to E-W trend (Stäblein, 1968; Illies and Greiner, 1976). The kinematics of these neotectonic faults is mainly extensional. Studies of exploration wells at the Soultz-sous-Forêts geothermal reservoir site, located near the WBF in the middle URG, revealed an active fracture set in the basement rocks with conjugate, N-S striking fractures and high dips (Dezayes et al., 2004). The good correlation in fault orientation and faulting mechanism of the trench site with the neotectonic faults and the basement fractures of the region imply that the structures at the trench site are part of an active tectonic fault system.

2.5.4. Surface trace of the fault system

Analyses of structural data strongly suggest that the fault zones observed in the trenches are part of the same fault system. In plan view, a line drawn between the fault zones in TR1 and TR3 has a trend of 020° (Fig. 2.4A). The geophysical measurements from the trench site give additional information on the trace of the fault system. Geophysical profiles taken parallel to TR1 and TR3 exhibit deformed reflection patterns and a distinct inclined resistivity contrast at the same location as the fault zones in the trenches. It is therefore concluded that the structures revealed by these measurements (Fig. 2.5) represent indeed the fault zone. A single straight line passing through all known fault zones determined by geophysical measurements and trenching cannot be constructed. Consequently, it is suggested that several overlapping and merging fault zone segments exist, yielding a general strike for the fault system of 015° (NNE-SSW, Fig. 2.4A). This strike is in good agreement with the Rhenish trend of the WBF. Based on the similarity in strike and on the location of the trench site, the fault system seen in the trenches and the measurements is regarded as a near-surface expression of the WBF. At the trench site, this near-surface expression is defined as a complex fault system consisting of several segments and extending from the base of the scarp to V1 over an area several tens of meters wide (Fig. 2.4A).

2.6. Reconstruction of trenches 1, 2 and 3

The excavation of the trenches provided an insight into the near-surface expression of an active fault system and the sedimentary conditions at the scarp. In order to obtain information on the chronological order of tectonic, erosional and depositional events at the site, a reconstruction has been built by backstripping along two sections: section A with TR1 and TR2 and section B with TR3 (Fig. 2.11, see also Fig. 2.4B). Eight evolutionary steps are included. The timing of these steps has been determined using absolute ages from TL and relative ages from correlation with marine isotope stages (MIS; ages after Shackleton, 1987). Since only few absolute ages are available, the correlation with MIS is incorporated in order to support and present as complete a reconstruction as possible. The correlation assumes the general simplifying concept that interglacial periods coincide with river incision whereas glacial periods coincide with fluvial accumulation (e.g. Penck, 1910; Büdel, 1977). It should be noted that recent insights show that the fluvial response to climate change is complex in detail (Van Balen et al., 2003).

Step 1 – erosion and deposition from Late Würm to Present (~ 12 ka – Present)

The present-day situation is documented by 3 trenches, numerous cores and geophysical measurements (Fig. 2.11, step 1). A soil and colluvium cover accumulated to more than 2 m on parts of the trench site and smoothed the surface. Preceding human activity caused partly removal of units 1B, 3B and 4B and created a steep slope near the scarp base. Step 1 includes the historic and prehistoric period. Late Middle Age artifacts in the lowest colluvium horizons give a maximum age for the colluvium accumulation (Table 2.1). TL dating yields a Late Würm age for the deposition of unit 4B.

Step 2 – faulting activity between Late Würm and Early Holocene (19 – 8 ka)

Step 2 represents the faulting event (Fig. 2.11, step 2). The fault zone in section A 'concentrates' at the base of the scarp. Since TR1 uncovers only the eastern part of the fault zone, it is suggested that the fault zone extends further west into the scarp (Fig. 2.11, step 2, section A, dashed lines with question marks). In TR3, a well defined,

5 m wide fault zone occurs, which is shown as a narrow zone in section B. The deformation of unit 4B in this section is an interpretation. It is concluded from section A that this unit also covered the slope at section B during step 2 and has been eroded between step 1 and 2. A GPR profile from the base of the scarp in section B (GPR7) shows a similar reflection pattern as the profile at TR1 (GPR8) and suggests that another fault zone may exist at this location (Figs. 2.5C and 2.11, step 2 dashed line with question mark). The faulting event occurred after the deposition of unit 4B and before the formation of the Holocene soil, unit 5B. TL dating from TR1 yields a maximum age of 19 ka and a minimum age of 8 ka for the event (Table 2.1).

Step 3 – deposition of unit 4B in the Late Würm (23 – 14 ka)

Prior to the faulting, unit 4B accumulated on the slope and smoothed the scarp (Fig. 2.11, step 3). Eastward dipping layers and strata diverging towards the east point to a downslope transport of the material. TL dates of 23 – 14 ka give a Late Würm age for the deposition of unit 4B (Fig. 2.11, step 3; Table 2.1).

Step 4 – deposition of unit 4A in the Middle to Late Würm (31 – 22 ka)

It is inferred from TR1 in section A that unit 4A was deposited on both the slope and at the base of the scarp (Fig. 2.11, step 4). For section B, no information from the base of the scarp exists, but a similar situation to section A has been assumed for the reconstruction. TL dates of 31 – 22 ka give Middle to Late Würm age for the deposition of unit 4A.

Step 5 – erosion and retreat of the scarp in the Middle Würm (37 – 31 ka)

The onlap of unit 4A onto unit 3B in TR1 indicates that unit 3B is older (Fig. 2.11, step 4). Furthermore, it suggests that unit 3B originally extended eastwards and that parts of the unit have been eroded. Step 5 represents this erosional phase. TL dating of units 3B and 4A suggests that the erosion took place between 37 – 31 ka (Fig. 2.11, step 5; Table 2.1). The erosion caused a retreat of the scarp and terminated at the location of the present-day fault zone in TR1 (Fig. 2.11, step 1). It seems unlikely that this lateral termination is coincidentally located at the present-day fault zone. It is possible that a fractured material influenced the erosion, thus suggesting penultimate faulting during step 5. The faults excavated in TR1 don't show offsets of this event but other faults that have not been excavated could exist and be related to the event (Fig.

2.11, step 5, dashed lines with question marks).

Step 6 – deposition on the slope from Early to Middle Würm (~ 100 – 37 ka)

At step 6, units 3A and 3B covered the scarp and created a smooth surface (Fig. 2.11, step 6). Dating of units 3A and 3B is difficult since a sample from the base of unit 3B delivers an age of > 100 ka (Fig. 2.11, step 6; Table 2.1) suggesting that unit 3A is older than 100 ka. The correlation of sedimentary events with MIS suggests to a cold period for the deposition of the units 3A and 3B. From MIS 5b (~ 100 ka) the climate was gradually cooling and reached a minimum during the MIS 4 (74 – 59 ka). This would suggest that the most likely rapid deposition of unit 3A initiated at ~ 100 ka and was followed by the deposition of unit 3B which lasted until 37 ka (step 5).

Step 7 – erosion and initial formation of the scarp during Eemian (~ 130 – ~ 100 ka)

The base of units 3A and 3B show the limit of erosion into the older deposits (Fig. 2.11, step 7). This lateral and vertical erosion caused the initial formation of the scarp at the trench site. Similar to step 5, it is possible that faulting activity influenced the position of this initial scarp. The correlation with MIS infers a warm period for erosion, probably the MIS 5e (Eemian interglacial). Considering that the deposition of unit 3A started in the MIS 5b (step 6) the erosional event at step 7 may have occurred between MIS 5e until MIS 5b (~ 130 – ~ 100 ka).

Step 8 – deposition of a fluvial terrace and soil formation during Riss and Eemian (~ 450 – ~ 130 ka)

The final step of the reconstruction suggests that a flat surface existed and that a soil, unit 2B, developed on the cross-bedded sands of unit 2A, (Fig. 2.11, step 8). The large thickness (2 m) of this soil observed in TR3, section B, implies that it extended over the whole trench site area. It is concluded that the 2B soil has been eroded at section A, most likely during step 7. The underlying units (1A, 1B, 2A) were deposited on a plain. Unit 2B and the sands of unit 2A are regarded as forming a terrace body. This terrace is located about 20 m above the Würm lower terrace, located at the base of the scarp, and could therefore belong to the Middle Terraces from the Riss glacial period (~ 350 – 130 ka). The thickness of unit 2B implies a long soil formation process, which could have initiated during the last interglacial (Eemian, MIS 5e, ~ 130 ka).

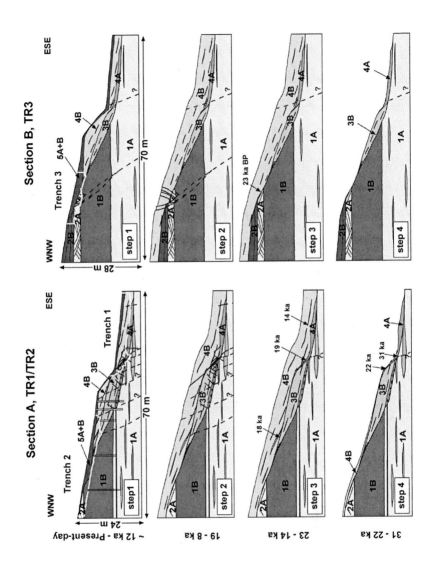

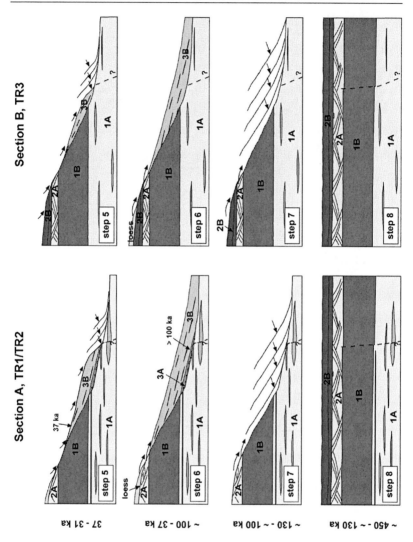

Figure 2.11. Reconstruction of the sequence of events at the trench site along two sections shown on Figure 2.4B. Section A follows the trace of TR1 and TR2. Section B includes TR3 and the location of GPR 8 profile. Based on the GPR 8 profile (Fig. 2.5C) section B suggests another fault zone at the base of the scarp (dashed line with question mark).

2.7. Discussion

2.7.1. Discussion on the origin of faults

The results of trenching and geophysical measurements show that the region exhibits near-surface fault structures. However, the presence of near-surface fault structures does not reveal information on their origin as different processes can cause surface faulting similar in style to the structures observed in the trenches. The trench site is situated along the border fault of an active rift system. Thus, the faults observed may have a tectonic origin. However, both surface processes and tectonic faulting can create slopes, which could be large enough to generate mass movements and could result in near-surface fault structures. This leads to the following questions: are the faults at the trench site of tectonic or gravitational origin, and if the origin is tectonic, was the movement aseismic or seismic (i.e. creeping or rapid)?

The structural analysis of the faults in the trenches revealed a consistent style and orientation, which correlates well with independent data from regional trends of active faults (Straub, 1962; Stäblein, 1968; Anderle, 1974; Schneider and Schneider, 1975; Illies and Greiner, 1976; Dezayes et al., 2004). This supports their tectonic origin. If however the faults in the trenches have a gravitational origin, one or more large slides would be required. This appears unlikely since no morphological indication of large slides is observed at the trench site. Nevertheless, small mass movements did occur at the site, as evidenced by the folds and decollements in TR1. Another aspect to consider is the position of the fault zone in the trenches relative to the lithological units. The fault zones are both located at the base of the scarp within slope wash and sands (units 1A, 3B, 4A, 4B in TR1) and on top of the slope within subhorizontally bedded units (units 1B, 2A; TR3). A slide generating such faults would be initiated on the sloping lithological boundary between the impermeable unit 1B and the overlying units. However, along this particular contact no features indicative of sliding were observed.

Due to the setting of the trench site, e.g. slope and Late Würm permafrost conditions, a non-tectonic origin cannot be entirely excluded. However, based on all observations, i.e. the correlation with regional fault trends, the consistency in deformation style across the trenches and the lack of mass movement features at the most prone position, the faults in the trenches have mainly a tectonic origin. In order to gather constraints on the tectonic strain rate, i.e. seismic or aseismic deformation, the present deformation pattern has to reveal temporal information about the faulting process. Since numerous analogue models show that the generation of similar fractures in sandy material both takes place at rapid and creeping rates (e.g. McClay and Ellis, 1987; Vendeville et al., 1987; Davy and Cobbold, 1991), the brittle deformation of the sandy units exposed in the trenches cannot reveal temporal information. In contrast, the clay-rich material of the 11 m thick unit 1B observed in TR3 appears to have a time-dependent rheological behavior, with the brittle deformation of this unit likely to be related to relative high strain rates. To prove whether the deformation occurred at high or low strain rates, the exact composition and conditions under which the clay-rich material deformed (e.g. degree of water saturation) would have to be known. Unfortunately, this information is unknown and cannot be retrieved. However, the liquefaction features observed in TR1 could be an indication of co-seismic deformation. Obermeier (1996a) developed five criteria to prove earthquake-induced liquefaction: (1) Rapid injection of material under overpressure, (2) the existence of several types of features, (3) ground water settings, (4) occurrence at multiple locations and (5) formation in short, discrete episodes. Criteria 1, 2 and 3 clearly apply to the trench site. No data is available for criteria 4 since the intensive land use of the study area precludes a regional study of liquefaction features. Furthermore, criteria 5 cannot be proven because the features in TR1 cannot be dated with sufficient precision. In addition to the 5 criteria, gravitational sliding has to be ruled out for inducing liquefaction (Obermeier, 1996a). The most prominent liquefaction features occur in the footwall of TR1 within clay-rich and cohesionless sand (Fig. 2.7). Since the lithological units affected remained in-situ and show no sign of mass movement, these features are regarded as indicating a co-seismic event. For the liquefaction features observed in the hangingwall (Figs. 2.6A, 2.6E) a relation to mass movements cannot be excluded. Based on these arguments, the liquefaction features observed in TR1 are interpreted to be earthquake-induced.

2.7.2. Paleoseismological implications of this study

Since it remains unclear whether the tectonic deformation observed in the trenches was related to continuous creep or to a single event, two scenarios for the formation of the maximum vertical displacement observed are possible. If the fault displacement is the result of creep, the minimum average fault displacement rate is 0.04 mm/yr (0.5 m / 19 – 8 ka). If the fault displacement was caused by a single seismic event with a minimum vertical displacement of 0.5 m, calculation of an earthquake magnitude yields a moment magnitude of $M_w = 6.5$ (using Wells and Coppersmith, 1994) empirical formula).

A creep rate of 0.04 mm/yr is relatively low, though it is of the same order of magnitude as average displacement rates of the region. The Quaternary uplift rate of the Mainz Basin obtained from the northern part of the WBF scarp (170 m uplift at Oppenheim; Brüning, 1977) is ~ 0.09 mm/yr. A paleoseismic event of M_w 6.5 is larger than the low-magnitude historical earthquakes observed in the northern URG. Unfortunately, detailed information on multiple seismic events and their timing cannot be derived from the trench site and the recurrence interval for large earthquakes cannot be constrained. Additional data from paleoseismological trenching would be necessary to support the observations of a possible M > 6 event of this study.

2.7.2.1. Recurrence intervals obtained from paleoseismology and seismology

For a comparison of the range of recurrence intervals for M > 6 earthquakes, that might be expected for the northern URG, estimates from trenching studies in the central URG, the Basel area and the Lower Rhine Graben could be considered. Table 2.2 summarizes several paleoearthquakes and their characteristics identified in trenches. The list demonstrates that for the URG and the Lower Rhine Graben (LRG) recurrence intervals for M > 6 earthquakes are in the order of tens of thousands of years, whereas the Basel area exhibits an exceptionally high recurrence interval of a few thousand years only. Ahorner (2001) investigated the seismicity of the LRG. Based on magnitude-frequency relationships this author derived recurrence intervals for the Erft fault system, a major fault system that has been active throughout the entire Tertiary and Quaternary. For a fault segment length of 13 km, which is considered a reasonable length for a fault rupture in this area, the recurrence interval for an M 6 earthquake is in the order of tens of thousands of years (Table 2.3). Schmedes et al. (2005) present magnitude-frequency relationships including paleoseismic data for the LRG. These authors demonstrate the importance of including this data, in particular for deriving a better estimate of the upper bound magnitude. To date, a thorough study of this kind has not been performed for the URG. At this stage, the recurrence interval for an M > 6 earthquake for the URG can only be estimated on the basis of the available studies (Table 2.2). Comparing the range of recurrence intervals for the URG and LRG the recurrence interval for such an earthquake in the northern URG is most likely in the order of tens of thousands of years.

Region	Fault investigated	Number of paleoearthquakes identified	Estimated magnitude [M_w]	Average displacement rate [mm/yr]	Average recurrence interval
URG	Western Border (this study)	1 in 8,000 to 19,000 years	6.0 – 6.5	0.04	–
	Achenheim-Hangenbieten	Not given (1), according to recurrence interval given 12 events in 300,000 years	6.0 – 6.5	0.03 (Achenheim), 0.04 (Hangenbieten)	25,000
Basel	Basel-Reinach	3 between 6480 BC and 1356 AD (2)	6.4 – 6.5	0.21	1,500-2,500
LRG	Feldbiss / Geleen	5 in 44,000 years, last event around 600-900 AD (3), activity before 15,000 years (4)	6.2 – 6.4 (5)	0.018 – 0.1	15,000 for trenches 1-3, 29,000-38,000 for trench 4
	Rurrand	3 in 40,000 years, last event around 400-1670 AD (6), suggest presence of coseismic event but unambiguous proof (7)	6.8 for last event (6)	0.05 – 0.2 (6)	–
	Peel	2 between 13,000-15,800 (8)	6.0 – 6.6	–	–

Table 2.2. Summary of identified paleoearthquakes and their characteristics in the URG and Lower Rhine Graben (LRG). (1) Cushing et al. (2000), (2) Meghraoui et al. (2001), (3) Vanneste et al. (1999), Meghraoui et al. (2000), Vanneste et al. (2001), (4) Houtgast et al. (2003), Houtgast et al. (2005), (5) Meghraoui et al. (2000), (6) Vanneste and Verbeeck (2001), (7) Lehmann et al. (2001), (8) Van den Berg et al. (2002).

Location	Magnitude [M_w]	Recurrence interval	
		50 km length of fault system	13 km length of fault system
Erft fault system	5.5	800	3,900
	6.0	2,200	11,000
	6.3	4,900	24,000

Table 2.3. Recurrence interval for fault segments of the Erft fault system, Lower Rhine Graben. Data after Ahorner (2001).

2.7.3. Discussion on the origin of the scarp

The sequential reconstruction of the trench site shows that, in addition to tectonic activity, fluvial dynamics have played a major role in the evolution of this part of the scarp (Fig. 2.11). Step 5 of the reconstruction of the trenches shows that lateral erosion terminates within the sandy and silty unit 3B and at the location of the present-day fault zone (Fig. 2.11). This coincidence cannot be explained by the mechanical behavior of the material and there is no reason why the erosion would not have extended westward into the more resistant clay-rich unit 1B. One possible explanation for this limit of erosion could be a fractured, and thus weakened, material (unit 3B) whereas the material further west was intact, and thus stronger. Conversely, the existence of an erosional scarp (formed during steps 5 and 7) could also have influenced the location of the surface termination of the fault zone during a later faulting event. It remains an open question for the trench site whether an erosional topography influenced faulting or pre-existing faults influenced erosion. At the northern end of the scarp, the influence of fracturing on erosion can be inferred for historic times. Here, it has been suggested that pre-existing faults subsidiary to the WBF have aided large landslides (Steuer, 1911; Fig. 2.2) by strongly influencing the permeability on the slope.

The interplay of fluvial dynamics and tectonics is also supported by the larger scale geomorphology of the area. The margins of the Mainz Bingen Graben are formed by a sequence of fluvial terraces, i.e. wide horizontal surfaces on top of the plateau (Early Pleistocene terraces), multiple small terraces along the scarp fronts (Middle to Late Pleistocene terraces) and the wide surface of the Late Pleistocene Lower terrace at the base (Kandler, 1970; Abele, 1977). The morphology of the WBF scarp closely resembles these margins (Figs. 2.2, 2.12). It is thus suggested that the wide surfaces on top of the plateau and the smaller surfaces on the WBF scarp front are fluvial terraces of the River Rhine or a tributary. The arrangement of the terraces implies an eastward migration of the river (Fig. 2.12). The suggestion that fluvial terraces form the scarp front is supported by the trench site interpretation (Fig. 2.11, step 8).

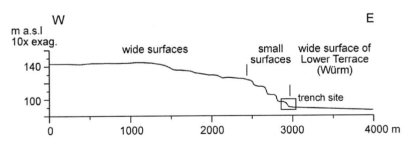

Figure 2.12. West to east topographic profile crossing the scarp at the trench site shown on Figure 2.3. The profile is based on digital topography data (DGM5, 2002) and field observations and shows a succession of wide and small horizontal surfaces interpreted as fluvial terraces.

The terraces of the Mainz Bingen Graben are interpreted to have formed in response to regional tectonic uplift that also caused the reactivation of faults along the Mainz Bingen Graben and the migration of the River Rhine away from the Mainz Basin (Wagner, 1930; Kandler, 1970; Abele, 1977). The terraces along the WBF scarp may have also been influenced by this regional uplift. The formation of the scarp would then be the result of fluvial erosion under increased regional tectonic activity. In addition, the concentration of multiple small terraces along the scarp front coinciding with the position of the WBF suggests a strong relationship between terrace and scarp formation and localized fault activity. Brüstle (2002) suggests a similar relationship for the Rhine River Fault scarp at the southern end of the URG (Fig. 2.1A). Recent studies on the Frankenthaler Terrace scarp (a lower terrace of the River Rhine; Stäblein, 1968), 20 km south of the study area, also suggest a tectonic control on the scarp formation, since the scarp coincides with faults mapped in the subsurface (Weidenfeller and Kärcher, 2004).

2.8. Conclusions

In summary, the integration of shallow geophysics, paleoseismological trenching and structural analysis enabled to locate and characterize near-surface deformation structures of the WBF at the base of the southern end of the scarp. The results of 4 trenches point to extensional faulting producing a consistent, conjugate fault set of 015°-striking faults, parallel to the WBF strike, and maximum vertical displacements in the order of 0.5 m. TL dating of deformed sediments shows that the faulting event occurred between 19 and 8 ka. The faulting may have been caused by a single seismic event with a moment magnitude of 6.5 and/or by creep movements. Reconstruction of the sequence of events at the trench site shows the local scale interplay between tectonic activity on the WBF and fluvial and erosional processes. This interplay is also important at a regional scale. Thus, the 20 km long WBF scarp, which was associated with the interplay of fluvial dynamics, erosion, regional uplift and localized activity on the WBF, has a mixed origin.

CHAPTER 3

PLEISTOCENE TECTONICS INFERRED FROM FLUVIAL TERRACES OF THE NORTHERN UPPER RHINE GRABEN[1]

3.1. Introduction

This study uses records of fluvial terraces in order to infer Pleistocene faulting activity. The area investigated is situated along the western margin of the northern URG and comprises additionally adjacent areas including the Mainz Basin and the southern part of the Rhenish Massif with the Upper Middle Rhine Valley (Fig. 3.1). The River Rhine flows across these structural units and the major faults separating them: the Western Border Fault (WBF) of the URG and the Hunsrück-Taunus Boundary Fault (HTBF). At the regional scale, the URG serves as a depocenter for the sediment load of the River Rhine and its tributaries. Since the Early Miocene (Middle Burdigalian; Fig. 1.2), the depositional environment in the URG is fluvial. The dominance of the northern part of the URG as a depocenter during Quaternary times is illustrated by the accumulation of 350 m and possibly more Quaternary sediments in the *Heidelberger Loch* (Bartz, 1974). This depocentre, which is associated with the eastern graben border fault, displays the geometry of an asymmetric graben. Whilst the eastern part of the URG was subsiding, its western parts experienced uplift and fluvial erosion, which resulted in the development of the ridges and valleys of the Vorderpfalz (Fig. 3.1). These ridges are mainly built up of Pliocene fluvial sands and Pleistocene fluvial terrace deposits, that are covered by Upper Pleistocene loess (Stäblein, 1968). Since the Quaternary, landforming processes dominate also in the Mainz Basin. The rivers began to incise and a tableland landscape evolved (Brüning, 1977). Regional uplift of the Mainz Basin led to the formation of a sequence of terraces, which were

---

[1] This chapter is mainly based on the publication: Peters, G., Van Balen, R.T., 2007. Pleistocene tectonics inferred from fluvial terraces of the northern Upper Rhine Graben, Germany. Tectonophysics 430: 41-65.

covered with loess during Late Pleistocene times (Kandler, 1970). Further to the northwest in the Rhenish Massif, Quaternary uplift was largest and yielded the narrow Middle Rhine Valley with a well-developed staircase of Rhine terraces (Bibus and Semmel, 1977; Meyer and Stets, 1998; 2002).

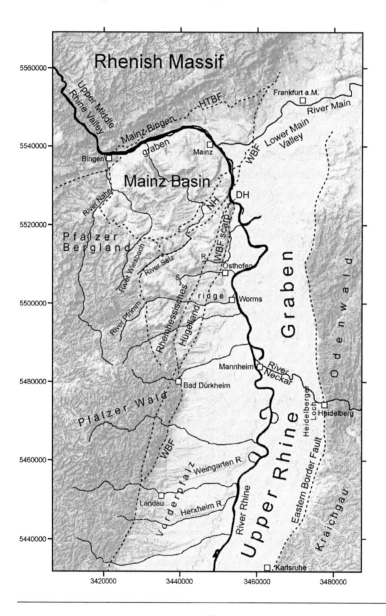

Figure 3.1. Geographical map of the study area. Previous terrace studies focused on small regions within the study area, namely the Vorderpfalz, the River Pfrimm Valley, the Lower Main Valley, the Mainz Bingen Graben and the Upper Middle Rhine Valley. Main faults are indicated with dashed lines. The location of several ridges is shown: the ridge north of the River Pfrimm Valley and the Weingarten and Herxheim ridges. R = Riederbach, S = Seebach. The shaded relief maps in Figures 3.1 and 3.2 were created with SRTM data (data source see caption of Figure 1.4). Coordinates of all maps are in the German Gauss-Krüger coordinate system.

To date, the characteristics of neotectonic activity in the northern URG and surroundings are not well constrained because little emphasis has been placed on mapping young faults. This lack of data can partly be explained by the setting. The study area is characterized by intensive modifications of the landscape due to high rates of Quaternary surface processes and human modifications. Therefore, the records of relatively slow tectonic deformation are less well preserved and thus difficult to detect. The seismicity of the study area documented for the last 1200 years is low with few damaging earthquakes of intensity VII (Leydecker, 2005a). This means that the potential for large, damaging earthquakes, with typical recurrence intervals of tens of thousands of years for intra-plate settings, and the seismic potential of individual faults cannot be established using the relatively short historic seismic catalogue alone. Given this seismo-tectonic setting, the investigation of geomorphological records of fault movements can be used to derive data on the long-term tectonic deformation of an area and the behavior of individual faults.

This study uses the relative position (elevation) of Pleistocene fluvial terraces to determine the more local effects of fault movements on terraces. The term "fluvial terrace" used in this study encompasses both landform and sediments. Where a distinction is needed, the landform is referred to as terrace surface and the sediments as terrace deposits. Terrace surfaces are interpreted to document the position of former valley floors. By correlating surfaces that have the same age, a longitudinal profile of the former valley floor can be reconstructed. Based on such a profile the effect of fault movements on terraces can be determined and displacement rates and patterns can be evaluated. For the reconstruction of a profile ranging from the western margin of the northern URG to the Upper Middle Rhine Valley, data from previous studies is included. It is supplemented with new terrace mapping based on field observations and topographic map interpretations. The newly mapped terrace surfaces are situated along the WBF at the border of the Mainz Basin and are correlated based on their morphology to previously mapped terraces in the vicinity. This correlation will allow discussing the transition from the subsiding northern URG to the uplifted Rhenish Massif.

Movements and displacement rates presented here for a new longitudinal profile strongly depend on the correlation scheme used. The mapping of terraces is relatively well established at the northwestern end of the URG and in the Middle Rhine Valley (e.g. Kandler, 1970; Bibus and Semmel, 1977). However, due to the lack of absolute ages of terrace deposits determination of a common terrace stratigraphy is problematic. This leads to uncertainties in the correlation of terraces between different regions. The problems in regional terrace correlation and the need for dating terraces in the study area have been frequently discussed in the past (e.g. Birkenhauer, 1973; Quitzow, 1974; Bibus and Semmel, 1977; Semmel, 1983; Hoselmann, 1996; Meyer and Stets, 1996). The lack of absolute ages of river terraces is also problematic in other Central European areas. Recent dating results of River Danube terraces in Hungary revealed unexpectedly young ages and demonstrated how unreliable previous age estimates have been (Ruszkiczay-Rüdiger et al., 2005a; Ruszkiczay-Rüdiger et al., 2005b). This study also demonstrates the need to improve the dating of terraces. The focus of this study lies on the determination of Pleistocene faulting activity. In this context, the correlation between terraces of the URG and the Rhenish Massif proposed herein and the fault movements inferred from this correlation scheme can, despite the inherent uncertainties, provide a first order quantification and additional constraint on the characteristics of the Pleistocene fault activity.

3.1.1. Main tectonic features in the region of Pleistocene fluvial terraces

In the Mainz Basin, several tectonic features are of relevance to this terrace study. The Niersteiner Horst is situated in the northeastern part of the Mainz Basin (Fig. 3.2A) in the northern continuation of the Permian Pfälzer anticline. Before and during the Tertiary, this horst represented a structural high. The relatively few Tertiary sediments that were deposited on the horst have been eroded during uplift of the structure. It is estimated that an uplift of 160 m has occurred since Early Pliocene (Sonne, 1972). Pleistocene uplift of the horst exposed Permian sediments and increased the NW tilting of these sediments (Sonne, 1969). The Pliocene-Pleistocene uplift of the Niersteiner Horst is considered rapid. It caused a morphological division of the Mainz

Basin into eastern and western plateaus (Brüning, 1977) and migration of the River Selz northwestwards (Klaer, 1977). Sonne (1969; 1972) suggested that recent tectonic activity concentrated on the southern boundary fault of the Niersteiner Horst, which joins the WBF north of the River Rhine, and the adjacent small structure of the so-called Dexheimer Horst (Figs. 3.1, 3.2A inset).

To the northwest of the Mainz Basin lies the so-called Mainz Bingen Graben (after Wagner, 1930). This small graben structure is located between the Rhenish Massif in the north and the Mainz Basin in the south. It is associated with the HTBF and has the same strike (Fig. 3.1). This latter fault is a Variscan terrane boundary that has been reactivated during formation of Late Variscan Saar Nahe Basin as well as during URG formation (Anderle, 1987). The HTBF is a folded fault dipping northwards at shallow depths and southwards at greater depths (profile in Figure 3.2B). Presently, seismicity concentrates along the HTBF although at a relatively low level (maximum intensity I_0 reached was V). The earthquake focal mechanisms indicate extensional movements of the fault at greater depths (Ahorner and Murawski, 1975).

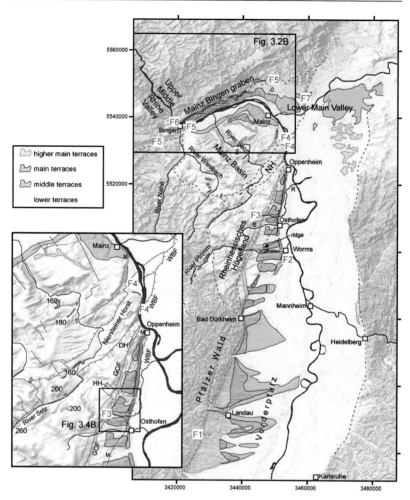

Figure 3.2. A) Map of fluvial terraces in the northern URG, Mainz Basin and Rhine-Main area.
New terrace mapping of this study covers the region from the Rheinhessische Hügelland to the
WBF scarp. Distribution of terraces in other areas is after the mapping of authors listed in the
caption of Table 3.1. The general classification and correlation of terraces are based on the
scheme shown in Table 3.1. F1 = Klingbach Valley Fault, F2 = Pfrimm Valley Fault, F3 =
Riederbach Valley Fault, F4 = Niersteiner Horst, F5 = HTBF, F6 = splay fault of HTBF, and
F7 = WBF and related faults of Semmel (1978). R = Riederbach, S = Seebach. Inset: detailed
terrace and fault map of the WBF scarp and the northeastern Mainz Basin with the Niersteiner
Horst. Three horst structures in this region expose Rotliegend deposits. Numbers indicate
elevations of the higher main and main terraces of the River Selz. Fault mapping after Sonne
(1972), Illies (1974a), Stapf (1988) and Franke (2001). DH = Dexheimer Horst, GOF =

Grünstadt – Oppenheim Fault, HH = Hillesheimer Horst, M = Mosbach Sands.

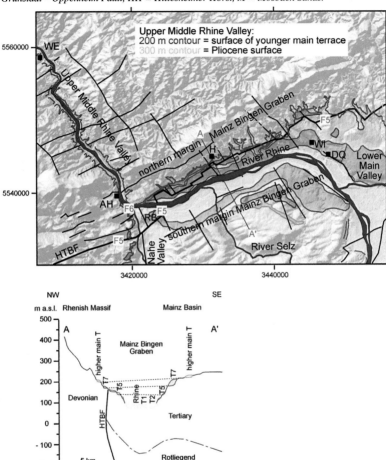

Figure 3.2. B) Map of fluvial terraces in the Mainz Bingen Graben (Kandler, 1970), Nahe Valley (Andres and Preuss, 1983) and Upper Middle Rhine Valley (Bibus and Semmel, 1977). After morphological correlation the 200 m contour corresponds to elevations of the main terrace surface (T7, t_{R5} and younger main terrace of River Nahe). The 300 m contour corresponds to the elevation of the Pliocene surface. Profile A – A' (modified after Figure 20 in Rothausen and Sonne, 1984) shows that significant displacements of the Hunsrück-Taunus Boundary Fault (HTBF, F5) are not noticeable in the morphology. Paleomagnetic measurements of Mosbach Sands were undertaken near Werlau (WE) in the Upper Middle Rhine Valley and in the Dykerhoff quarry (DQ) near Wiesbaden. Faults after Franke and Anderle (2001). AH = Assmannshausen, H = Hallgarten, RB = Rochusberg, WI = Wiesbaden.

3.2. Review of terrace studies

3.2.1. Overview

The following sections present a review of previous terrace studies, which focused on specific regions within the larger study area:

- the Vorderpfalz (Stäblein, 1968),
- the southeastern Mainz Basin along the River Pfrimm (Leser, 1967),
- the Mainz Basin along the rivers Wiesbach and Selz (Wagner, 1931b; 1935; Matthess, 1956; Sonne, 1956; 1972; Wagner, 1972; Franke, 1999; 2001),
- the Mainz Bingen Graben (Kandler, 1970),
- the Lower Main Valley (Semmel, 1969), and
- the Upper Middle Rhine Valley (Bibus and Semmel, 1977).

A detailed description of the stratigraphy of the individual terrace units mapped by the authors cited above is given in Tables A4.1 to A4.6 in the Appendix of this thesis (section A.4).

The current classification of terraces in the reviewed studies is not consistent. In order to compare terraces of the different regions and to synthesize the previous work, a new correlation is proposed in this study and presented in Table 3.1. Due to large uncertainties in the correlation of individual terraces (see section 3.4), a simplified terrace group hierarchy is used. This grouping is originally based on the classical subdivision of terraces in the Middle Rhine and Lower Rhine area (Kaiser, 1903) and is generally accepted among workers of the Rhine terraces. The simple hierarchy allows a first order correlation of terrace groups between different regions.

All authors cited in the following descriptions mapped the top of the terrace deposits. For most terraces in the study area, the top forms a planar surface that can be clearly identified in the morphology. The frequent covering of terrace deposits in the study area with a meter or more of loess or slope wash is mostly neglected, which means that the cover is added to the height of the terrace surfaces. In the existing studies, terrace ages are mainly relative and based on elevation of terrace surfaces. In some studies (Leser, 1967; Stäblein, 1968; Kandler, 1970), age determination is based on correlations with Quaternary glacial periods following the concept that glacial periods coincide with fluvial accumulation and build up of a terrace body (e.g. Penck, 1910; Büdel, 1977). Absolute dating of the lower terraces in the area has been performed using 14C and thermoluminescence and documents the Würm age of the terraces (e.g. Scheer, 1978); *Chapter 2 of this thesis*). For a unique middle terrace element, additional paleomagnetic measurements and paleontological records are available (Mosbach Sands; Kandler, 1970; Bibus and Semmel, 1977). Investigations of sedimentology and petrography are included in most of the previous studies, enabling correlation of terraces between the different regions (Table 3.1).

General classification	Vorder-pfalz	Pfrimm valley	Selz T	Wies-bach T	Rheinhess. Hügelland, WBF scarp (this study)	Lower Main Valley	Mainz Bingen Graben	General classification	Upper Middle Rhine Valley	Lower Middle Rhine Valley	Lower Rhenish Embay-ment	Chronostratigraphy and correlation with glacial periods	
Lower Terraces	Nieder-terrasse	Nieder-terrasse	Nieder-terrasse	Nieder-terrasse	Lower terrace	t7 / t6	T1 / T2	**Lower Terraces**	tR11 / tR10	YLT / oLT	ylT / oLT	Würm — 12 / 6-25 / 25-100	**Late Pleist.**
Middle Terraces						t5 / t4	T3	**Middle Terraces**	tR9	LMT 2	MT 6	Riss — 125-250	**Middle Pleistocene**
	Talweg-terrasse	Mittel-terrassen	Mittel-terrassen (120 – 147 m)	Mittel-terrassen (140 – 170 m)	Middle terraces	t3	T4		tR8 / tR7	LMT 1 MMT 2 MMT 1 UMT	MT 5 MT 4 MT 3 MT 2 MT 1	Mindel — 385-490	
									tR6	UT 4	UT 4		
						t2	T5		tR5 (jüngere Haupt-terrasse)	UT 2/3 (jüngere Haupt-terrasse)	UT 3 / UT 2	Cromer — 600-800	**Brunhes**
							T6 (Mosbach sands)		tR4 (ältere Haupt-terrasse)	UT 1	UT 1		
Main Terraces	Hoch-terrasse	Hoch-terrassen jüngere Haupt-terrasse (160 – 180 m)	Jüngere Haupt-terrasse (170 – 180 m)	Hoch-terrasse (170 – 180 m)	Main terraces	t1	T7a / T7	**Main Terraces**	tR3	Lower Pleisto-cene Terraces	Günz 800-1300 Ältest-pleistozän II — Donau 1300-1600		**Early Pleistocene**
Higher Main Terraces	Haupt-terrasse	ältere Haupt-terrasse	ältere Haupt-terrasse (180 – 260 m)	Haupt-terrassen 180 – 265 m	Higher Main terrace	Early Pleist. sands (younger Weisenau sands)	Ältest-pleistozäne Terrasse	**Higher Main Terraces**	tR2 / tR1		Ältest-pleistozän I	Bieber 1600-2600	**Matuyama**
Pliocene surfaces	Pliocene glacis surface					Pliocene sand (older Weisenau sands)		**Pliocene surfaces**			upper Pliocene	> 2600	**Gauss / Pliocene**

Table 3.1. Chronostratigraphic interpretation and correlation of terraces in the Upper, Middle and Lower Rhine Graben. Correlation between Mainz Bingen Graben and Upper Middle Rhine Valley after Bibus and Semmel (1977); between Upper and Lower Middle Rhine and Lower Rhine Graben after Boenigk and Frechen (2006). For the other areas a morphological correlation, which is established in this study, is used. Note that the main and middle terraces in the northern URG and Mainz Basin (left part of Table 3.1) do not coincide in age with the main and middle terraces in the Middle and Lower Rhine (right part of Table 3.1). This is discussed in detail in section 3.5.2. Terrace stratigraphy uses German or English terms from the authors cited as follows. Vorderpfalz (Stäblein, 1968); Pfrimm Valley (Leser, 1967); Selz

and Wiesbach terraces (Wagner, 1931b; 1935; Matthess, 1956; Sonne, 1956; 1972; Wagner, 1972; Franke, 1999; 2001) and this study; Lower Main Valley (Semmel, 1969); Mainz Bingen Graben (Kandler, 1970); Upper Middle Rhine Valley (Bibus and Semmel, 1977); Lower Middle Rhine Valley and Lower Rhine Graben (Boenigk and Frechen, 2006). Correlation with glacial periods and estimated ages after Kandler (1970), Scheer (1978) and Boenigk and Frechen (2006).

3.2.2. Vorderpfalz

Stäblein (1968) mapped the terraces of the Vorderpfalz (Fig. 3.2A). There, terraces slope to the east and are characterized by wide surfaces with small elevation changes. A simple classification scheme was used based on the relative elevation of terrace surfaces and on the content of bleached Buntsandstein gravel, which is assumed to decrease with decreasing age, i.e. lower terrace elevation, and with increasing distance to the Pfälzer Wald. In total, four terrace generations are proposed as, from oldest to youngest, Hauptterrasse, Hochterrasse, Talwegterrasse, Niederterrasse. These correspond to the higher main, main, middle and lower terraces in the regional classification (Table 3.1). Additionally, a Pliocene surface has been mapped, remnants of which occur in the westernmost part of the Vorderpfalz along the foothills of the Pfälzer Wald. Immediately to the east lie the subsequent Pleistocene terraces (Fig. 3.2A). All terraces older than the lower terrace were affected by erosion of rivers draining from the Pfälzer Wald so that several wide valleys dissect their surfaces. The terraces are at present preserved on several ridges (Fig. 3.2A). Due to the erosion, remnants of the middle terrace (Talwegterrasse) are narrow and are present as elongated surfaces at the rim of the main terraces. A loess cover of a few meters of thickness smoothes most terrace scarps in the Vorderpfalz (Stäblein, 1968). Only the middle terrace is separated from the older and younger terraces by a distinct scarp. The lower terrace covers the wide valleys of the Vorderpfalz. Stäblein (1968) interprets from the distribution of terraces that the River Rhine has formed the higher main and main terraces, whereas tributaries draining from the Pfälzer Wald have formed the middle and lower terraces.

3.2.3. River Pfrimm

A sequence of several small terraces along the course of the River Pfrimm has been mapped first by Weiler (1931) and subsequently by Leser (1967) using relative terrace elevations. The course of the River Pfrimm extends over three tectonic units:

the northern Pfälzer Wald, the southeastern part of the Mainz Basin and the western margin of the URG (Fig. 3.2A). At the junction of the Pfrimm Valley and the URG, the valley widens and terrace surfaces increase in size. Leser (1967) showed that during the period of the higher main terraces (Hauptterrassen), the course of the Pfrimm was towards the NE. With the onset of the main terraces (Hochterrassen) formation, the Pfrimm built a large fan at the entrance of the URG, a remnant of which is thought to be the ridge north of the Pfrimm Valley (Figs. 3.1, 3.2A). Subsequently, the Pfrimm changed its lower course to an E-W orientation and incised into the fan. This change in course is interpreted as a result of uplift of the southeastern Mainz Basin and subsidence in the URG accompanied by activity of an E-W oriented fault parallel to the valley of the lower Pfrimm course (Leser, 1967).

3.2.4. Rivers Selz and Wiesbach in the Mainz Basin

The two largest rivers draining the Mainz Basin are the Selz and Wiesbach (Figs. 3.1, 3.2A). Since terraces older than the higher main terraces (Hauptterrassen) have largely been eroded in the Mainz Basin (Brüning, 1977), only remnants of younger terraces are mapped (Fig. 3.2A). The terrace deposits are generally covered by loess and slope wash, and do not form planar surfaces. In addition, sliding of the slopes is a common phenomenon along the River Selz (Rogall and Schmitt, 2005), which could have reworked or covered the terrace deposits. Therefore, terrace mapping along the Selz and Wiesbach rivers is mainly based on coring (Wagner, 1931b; 1935; Matthess, 1956; Sonne, 1956; 1972; Wagner, 1972; Franke, 1999; 2001).

Along the River Selz, four terrace levels at different elevations can be distinguished. A uniform terrace classification for the entire River Selz course does not exist since previous studies focused on parts of the river course only. Wagner (1931b; 1972) proposes a classification in the lower and middle course of the Selz. Here, four terrace levels are distinguished: Talwegterrasse (120 – 147 m a.s.l.), jüngere Hauptterrasse (160 – 180 m), ältere Hauptterrasse (180 – 200 m) and ältere Pleistozäne Terrassen (220 – 260 m; Wagner, 1931b; 1972). Based on Wagner (1931b; 1972) the following classification according to the general scheme of this study is proposed (Table 3.1):

120 – 147 m = middle terraces

160 – 180 m = main terraces

180 – 200 m = higher main terraces (lower and middle course of Selz)

220 – 260 m = higher main terraces (upper course of Selz)

In the middle reaches of the Selz, no terraces have yet been identified. Instead, frequent remnants of so-called 'local gravels of unclear origin' occur at elevations between 140 – 170 m. They have been interpreted either as products of weathering transported locally by sliding (Sonne, 1972) or as local deposits of small tributaries transported short distances (Wagner, 1972). Alternatively, these gravels could also be remnants of main terraces due to their topographic position.

Along the River Wiesbach, few terrace remnants have been mapped in the upper and middle course (Wagner, 1935; Sonne, 1956; Franke, 1999). Based on current mapping and the classification of Wagner (1935) the following reclassification is proposed:

140 – 170 m = middle terraces (Talwegterrasse)

170 – 180 m = main terrace (Hochterrasse)

180 – 265 m = higher main terraces (Hauptterrasse)

3.2.5. Lower Main Valley

At the northeastern end of the URG, the River Main has formed a sequence of terraces in its lower course in the Lower Main Valley (Fig. 3.2A). Semmel (1969) distinguishes seven terraces (t1 to t7 from oldest to youngest; Table 3.1). The middle terraces of the River Main can be correlated with the middle terraces of the Rhine. The T3 of the River Rhine after Kandler (1970) equals the t4 and t5 of Semmel (1969) of the River Main (Bibus and Semmel, 1977; Table 3.1). The t1 and t2 formed due to aggradation of sediments, whereas t1 lies below t2. The younger terraces (t3 to t7) are related to down-cutting and aggradation phases and form a terrace staircase with the higher terraces being older than the lower ones. A morphological scarp is most distinct between the t2 and t3 terraces. This is interpreted as indication for increased fluvial incision before the t3 accumulation (Semmel, 1969).

3.2.6. Mainz Bingen Graben

A sequence of terraces occurs on the northern and southern side of the Mainz Bingen Graben with their ages being generally inferred from their relative elevations (first mapping Oestreich, 1909; Wagner, 1931b; detailed and revised mapping by Kandler, 1970; Figs. 3.2A, B). Kandler (1970) distinguishes a total of nine terraces, which can be grouped into one higher main terrace, two main terraces (T7, T7a), four middle terraces (T6 – T3) and two lower terraces (T2, T1; Table 3.1), almost all of which become narrower downstream. Remnants of the higher main terrace exist on both sides of the Rhine. On the northern side, they can be traced for almost the entire length of the Mainz Bingen Graben, whereas on the southern side only isolated remnants are present. The main terrace (T7) builds wide and nearly continuous surfaces on the northern side and few surfaces on the rim of the plateau on the southern side. An additional terrace level (T7a) below the main terrace (T7) occurs partly on the northern side of the graben. Kandler (1970) interprets this additional terrace to result from local movements along a NE-SW trending fault between Bingen and Hallgarten (Fig. 3.2B). Except for T6, the middle terraces (T6 – T3) are smaller than the older terraces. A major morphological scarp, 25 – 30 m high, separates the middle and main terraces. This scarp is the result of a down-cutting phase after the main terrace formation, which led to 50 m of incision and was followed by the accumulation of maximum 18 m of T6 deposits, also referred to as Mosbach Sands (Kandler, 1970). Both middle terraces T5 and T4 cut into and deposited onto the Mosbach Sands. Remnants of the T5 to T3 middle terraces are rare on the southern side of the Mainz Bingen Graben and more frequent on the northern side. The lower terrace forms wide surfaces on both sides of the River Rhine. Due to an erosional discontinuity, this terrace consists of two levels (T1, T2), only detectable with coring (Kandler, 1970; Sonne, 1977). The older level T2 is of Early Würm age. It is partly covered with Würm loess (Sonne, 1977). The younger terrace level T1 formed during Middle Würm. The bottom of T1 deposits lies up to 7 m below the present-day riverbed of the River Rhine (profile in Figure 3.2B). This shows that the river has not yet reached the erosional base of Middle Würm times (Kandler, 1970). A rise of the bottom of T1 from the eastern to the western Mainz Bingen Graben by 4 m has been detected in cores and is related to young fault movements (Kandler, 1970).

3.2.6.1. Mosbach Sands / T6 terrace

The Mosbach Sands are of special interest since they have been paleomagnetically dated and thus are used as a chronostratigraphic unit for regional correlation (e.g. Bibus and Semmel, 1977). Sands with the characteristic Mosbach facies have been described from numerous locations in the study area (Fig. 3.2A, inset):

- the northern URG (north of Pfrimm Valley near Abenheim, Semmel and Fromm, 1976; at Oppenheim, Schraft, 1979),
- the area of Wiesbaden at the eastern end of the Mainz Bingen Graben (Dyckerhoff quarry, e.g. Brüning, 1978),
- the Lower Main Valley (t2 terrace; Bibus and Semmel, 1977),
- the Upper Middle Rhine Valley (Bibus and Semmel, 1977) and
- the Lower Middle Rhine Valley (Hönninger Sands of Boenigk and Hoselmann, 1991).

In the Mainz Bingen Graben the Mosbach Sands have been strongly modified by later erosion and deposition of younger middle terraces, and therefore do not form a separate morphological terrace (Kandler, 1970). The sands contain a rich vertebrate fauna of both warm and cold periods (including 4 interglacials; Abele, 1977). Despite the wealth of data, the stratigraphic position of the sands remains unclear. This is because no interpretation is consistently supported by all available evidence (paleontological, pedological and climatological evidence; see Kandler, 1970). Paleomagnetic measurements revealed that the Mosbach Sands are older than the Matuyama/Brunhes boundary (780 ka), see below. Younger deposits in the upper part of the sands showed normal polarization of the Brunhes phase (Bibus and Semmel, 1977). Accepting the remaining uncertainties, the Mosbach Sands may still be used as a chronostratigraphical marker, particularly for the correlation of terraces in the Lower Main Valley with terraces in the Upper Middle Rhine Valley (Bibus and Semmel, 1977).

3.2.7. Upper Middle Rhine Valley

In the narrow Upper Middle Rhine Valley, the number of Rhine terraces increases. Bibus and Semmel (1977) distinguished eleven terraces (Table 3.1), which can be grouped into two higher main terraces ($t_{R1} - t_{R2}$), three main terraces ($t_{R3} - t_{R5}$), four middle terraces ($t_{R6} - t_{R9}$) and two lower terraces ($t_{R10} - t_{R11}$). The higher main and main terraces form wide surfaces above the narrow valley (Figs. 3.2B, 3.3). The middle terraces are located on the steep valley slope and form small surfaces (strath terraces) that are only partly preserved. The lower terraces are located on the valley floor on both sides of the present-day Rhine course. The height differences between the t_{R1} to t_{R6} terraces are relatively small with 5 to 15 m. A significantly large height difference lies between the t_{R6} and t_{R7} terraces with 45 m, while for the younger terraces the differences are again relatively uniform with 20 m (t_{R7} to t_{R9}, see also Figure 3.7). This situation leads to the interpretation that the t_{R6} terrace forms the transition between the main terraces with dominantly lateral erosion and the middle terraces with increased incision. During the t_{R6} period, incision was still minor while it reached a maximum before accumulation of t_{R7} (Meyer and Stets, 1996). Bibus and Semmel (1977) state that the t_{R4} terrace contains sands of the Mosbach facies. In addition, paleomagnetic measurements at an outcrop in the Upper Middle Rhine Valley (Werlau, see Figure 3.2B) revealed the Matuyama/Brunhes boundary within the t_{R4} deposits (in Bibus and Semmel, 1977, based on an unpublished report of Fromm, 1978). This data allows correlation with terraces in the Lower Main Valley and the Mainz Bingen Graben area (see Figure 3.2B for location of Mosbach Sands). Details of this correlation are discussed in sections 3.4 and 3.5.2.

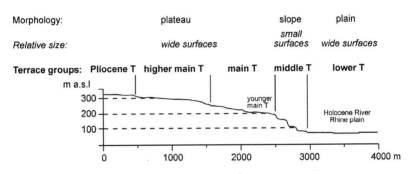

Figure 3.3. Schematic profile of Pleistocene terrace staircase in the URG and Middle Rhine Valley. The largest morphological step occurs between the main and uppermost middle terraces. The heights of the profile correspond to height levels of the Middle Rhine Valley. The vertical axis is 10 x exaggerated. In the URG, the topography is approx. 100 m lower and the slope is less steep.

3.3. New terrace mapping

3.3.1. Field mapping

The new terrace mapping of this study uses two methodologies: field surveys and topographic map analyses. The field surveys concentrate on the southern part of a 20 km long segment of the WBF that is situated in the northwest of the study area near the city of Worms (Figs. 3.4, 3.5 index map). For 20 km, the WBF follows the base of a linear 50 – 100 m high morphological scarp, referred to as WBF scarp. The work of this thesis (*Chapter 2*) revealed in several trenches surface displacements in the order of 0.5 m of the WBF, indicating a partly tectonic origin for this scarp. Tectonic deformation prior to the deposition of the sediments exposed in the trenches could not be confirmed through trenching or geophysical measurements for this site (*Chapter 2*). However, the 50 m scarp height cannot be fully explained by 0.5 m of surface displacement. Würm terrace deposits of the lower terrace are mapped at the bottom of the scarp (Franke, 2001). This age could be verified in the trenches by thermoluminescence dating of the deposits (*Chapter 2*). Trenches situated on the slope of the scarp about 10 m above the lower terrace exposed terrace deposits of possible Riss age (*Chapter 2*). These findings led to the hypothesis that also the slope of the scarp and the plateau was formed by a sequence of fluvial terraces, comparable to the profile shown in Figure 3.3.

The slope of the scarp has been intensively modified for viniculture for several centuries. Thus, the present-day sequence of vineyard terraces on the slope does not necessarily correspond to fluvial terraces. In this situation, intensive coring and/or trenching would be required to map fluvial terraces on the slope, which could not be performed in the framework of this study. For this reason, the mapping of new terraces in the vicinity of the trench site was limited to morphological field mapping (without coring) on the plateau (Fig. 3.4). The plateau bounding the southern part of the scarp to the west is mainly covered by Würm loess (eolian and alluvial) and partly by sands of Pliocene or Quaternary age (Franke, 2001). Field surveys of this study revealed frequent gravels and few cobbles (up to 10 cm in diameter) on the plateau. The gravels are well rounded and consist of sandstone (Buntsandstein), limestone, milky quartz and porphyry. These findings clearly indicate a fluvial transport from the western graben shoulder (Pfälzer Wald rock assemblage) and from the Mainz Basin (Tertiary assemblage) and suggest the existence of fluvial terraces on the plateau.

Between Osthofen and Alsheim, the plateau is incised by several streams, which separate it into three ridges (Fig. 3.4B). The morphology of the three ridges was mapped in the field. Each ridge was mapped from a viewpoint situated on the opposite ridge. From these viewpoints, stepped surfaces on the ridges were visible and documented in sketches (Fig. 3.4A). The sizes of the surfaces were determined relatively. The sketches in Figure 3.4A display these surfaces not to scale. Using a large-scale topographic map (1:5000), positions of the individual surfaces were identified and their elevations were estimated (elevations in Figure 3.4A). The sketches show a fine division of surfaces, which are interpreted as stepped fluvial terrace surfaces based on their morphology and the findings of fluvial gravels. The three profiles are generally very similar, with wide surfaces at the crest and several intermediate-sized surfaces below the crest and towards the front of the plateau. All surfaces from the crest to the front of the plateau are 20 – 30 m lower on the southern ridge in comparison to the northern ridges. The most significant break in slope on the two northern ridges is located at ~ 170 m. On the southern ridge it is located ~ 28 m lower at ~ 142 m (arrows in Figure 3.4A). This elevation difference is assumed to result from the activity of a fault parallel to the Riederbach Valley.

3.3.2. Mapping using topographic data

Field mapping of the ridges served as a starting point for regional terrace mapping using topographic maps (Figs. 3.4B, 3.5). Initially, the surfaces identified in the field have been located on a 1:5000 topographic map and on profiles. Fine division of individual surfaces is not possible at this scale but the general division in wide, intermediate and small surfaces can be made. This coarse division is also detectable on a 1:50,000 map (Fig. 3.4B). The comparison of the field sketches and the surfaces identified on the 1:50,000 map shows that the wide to intermediate size surfaces can be identified on the map shown in Figure 3.4B. Due to the poorer resolution of the map, individual small-scale surfaces cannot be detected. For this reason, Figure 3.4B contains less individual surfaces than Figure 3.4A and the estimated elevations deviate slightly.

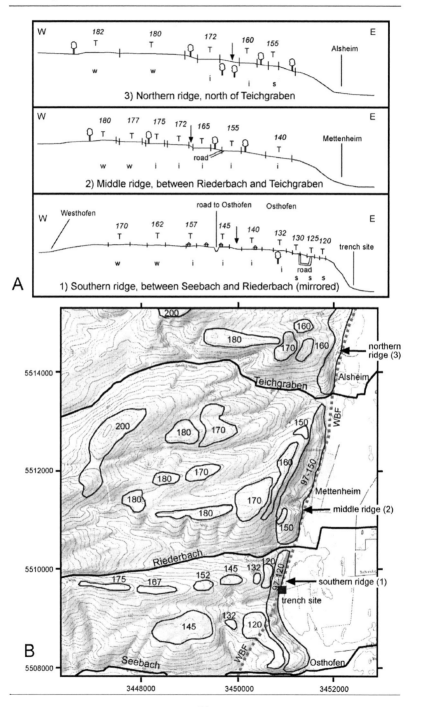

Figure 3.4. A) Cross-sections mapped in the field of three ridges of the plateau bounding the Western Border Fault (WBF). The WBF is situated at the base of the plateau (see also grey stippled line in Figure 3.4B). A sequence of surfaces, interpreted as terrace surfaces T, has been mapped. The most remarkable breaks in slopes are indicated with arrows. The surfaces vary in size (w = wide, i = intermediate, s = small) and lower in elevation from west to east. Elevations are taken from topographic maps of 1:5000 scale. B) Outline of terrace surfaces on the plateau. For location of the map see index map in Figure 3.5A. Individual surfaces are grouped according to the legend in Figure 3.5. To the north of the Riederbach Valley, terraces are 20 – 30 m higher than to the south of the valley. Source for topographic map: orohydrographic map 1:50,000, sheet L6314 O Alzey.

Using 1:50,000 topographic maps, terrace surfaces of wide to intermediate size have been mapped in the area south of the WBF scarp towards the Rheinhessische Hügelland and the Vorderpfalz (Figs. 3.5A, B), and along the WBF scarp towards Oppenheim in the north (Fig. 3.5C). The individual surfaces identified have been grouped and classified into lower, middle, main and higher main terraces (see legend in Figure 3.4). The terrace groups are characterized by individual surfaces of comparable relative sizes and equal elevations. The distribution of these groups in the entire region shows a narrow band of higher main terraces, a wide zone of main terraces and a narrow band of middle terraces (Figs. 3.2A, 3.5). From south to north, the width of each terrace group decreases.

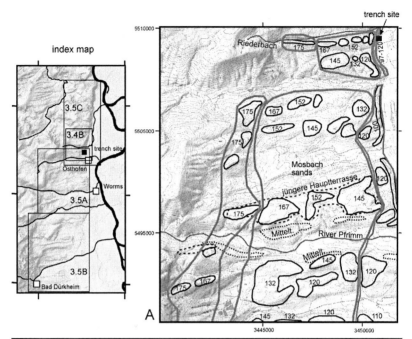

North of Riederbach	South of Riederbach	groups of terraces
97-180 middle terraces	97-120 middle terraces	*middle terraces*
150 higher middle terrace	120 higher middle terrace	
160 younger main terrace	132 younger main	
170 main terrace	145 main terrace	
	152 main terrace	*main terraces*
180 main terrace	167 main terrace	
200 higher main terrace	175 higher main terrace	*higher main terrace*

Figure 3.5. A) Topographic terrace mapping from the trench site southwards to the Rheinhessische Hügelland and the Pfrimm Valley mapped by Leser (1967). See index map for location of the individual maps. The elevations are in m above sea level. The legend shows the terrace groups, which are characterized by individual surfaces at equal elevations. Thick grey solid lines mark the extents of terrace groups. Terraces of the River Pfrimm by Leser (1967) are indicated with stippled lines. Mosbach Sands occur north of the Pfrimm Valley according to Semmel and Fromm (1976). Sources for topographic maps: A) orohydrographic map 1:50,000, sheet L6314 O Alzey, B) orohydrographic map 1:50,000, sheet L6514 O Bad Dürkheim, C) topographic map 1:100,000, sheet C6314 Mainz.

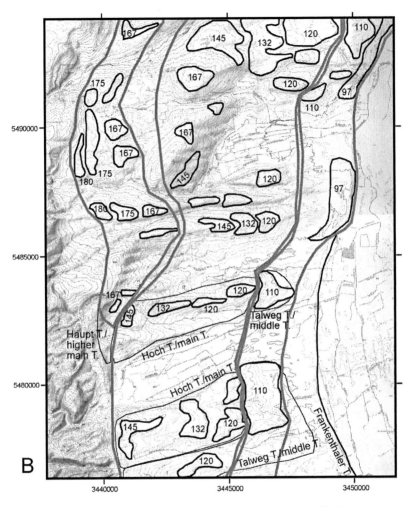

Figure 3.5. B) Topographic terrace mapping from the Rheinhessische Hügelland southwards to the Vorderpfalz and the area mapped by Stäblein (1968). Terraces of the Vorderpfalz are outlined with thin black lines.

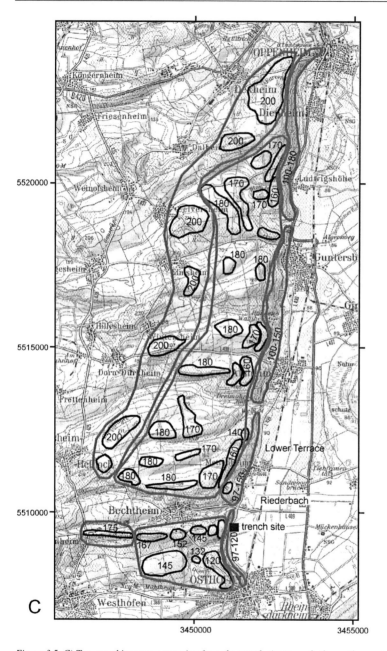

Figure 3.5. C) Topographic terrace mapping from the trench site towards the northern extends of the WBF scarp (Osthofen to Oppenheim).

3.4. Correlation of terraces

The terrace groups identified on 1:50,000 maps are correlated with the groups of previous terrace studies to the south and north (Vorderpfalz, Mainz Bingen Graben) with the goal of providing the first correlation of terraces for the western margin of the URG northwards to the Upper Middle Rhine Valley. Since the terrace mapping of this study does not provide stratigraphic constraints for the terraces, the correlation of these with terraces of previous studies in the vicinity is limited to the use of morphological criteria. These criteria include comparison of topographic elevation and relative sizes of the terraces. In order to avoid uncertainties in the mapping of individual terrace levels, the large-scale terrace groups are used for correlation.

3.4.1. Correlation of new terraces

The terrace mapping of this study extends southwards to the Vorderpfalz that was investigated by Stäblein (1968). At the overlap of the two study areas, a good correlation of elevations and relative sizes exists between the different terrace groups. The higher main, main and middle terraces of the new mapping correspond to the Hauptterrasse, Hochterrasse and Talwegterrasse of Stäblein (1968) respectively (Fig. 3.5B). Furthermore, a correlation between the newly mapped terraces and terraces of the Pfrimm Valley is also possible: the main terraces and the jüngere Hauptterrasse of the Pfrimm are equal in elevation and build the ridge north of the present-day River Pfrimm. It is thus suggested that both terrace units have the same age (Fig. 3.5A, Table 3.1). Likewise, the higher main terraces and the Hauptterrasse of the Pfrimm are most likely of the same age. The middle terraces of the Pfrimm are considered older than the middle terraces of this study due to their relative elevations (120 – 160 m a.s.l. of Pfrimm versus < 120 m a.s.l.).

Between the WBF scarp and the Mainz Bingen Graben lies the Niersteiner Horst, which experienced significant uplift during Pliocene and Quaternary times (Sonne, 1969; 1972; Fig. 3.2A, inset). Only terrace remnants have been preserved on the horst structure and its flanks (higher main terrace on the horst, Wagner, 1962; main terrace east of the horst at Oppenheim, 155 – 165 m a.s.l., Steuer, 1911; Kandler, 1970; Mosbach Sands at Oppenheim at c. 160 m a.s.l. after Schraft, 1979, indicated with M in Figure 3.2A, inset). Given this tectonic setting and the scarce data, the correlation of terraces between the WBF scarp and the Mainz Bingen Graben is limited to the use of morphological criteria. Both the WBF scarp and the terraced margins of the Mainz Bingen Graben exhibit wide terrace surfaces on the plateau and several small terraces on the slopes. It is therefore suggested that the higher main and main terraces form the plateau (higher main and T7 of Kandler) and the middle terraces form the slopes (T6 – T3 of Kandler). The correlation adopted implies that incision occurred simultaneously along the WBF scarp and the Mainz Bingen Graben (see discussion in section 3.5.2 and Figure 3.8).

3.4.2. Correlation between Mainz Bingen Graben and Upper Middle Rhine Valley

Downstream, the correlation between fluvial terraces of the Mainz Bingen Graben and the Upper Middle Rhine Valley across the Hunsrück-Taunus Boundary Fault (HTBF) leads to several uncertainties even though there is a wealth of data for both regions. So far, two scenarios have been proposed that are based on stratigraphic or morphological correlations of the terraces. The implications of these scenarios on the reconstruction of fault movements of the HTBF are discussed later (section 3.5.2). Based on these implications it is concluded that the stratigraphic correlation gives plausible results, whereas the morphologic correlation does not.

The stratigraphic correlation is based on the correlation of Mosbach Sands / T6 such that the t_{R4} terrace of the Upper Middle Rhine Valley and the t1 in the Lower Main Valley correspond to the Mosbach Sands / T6 in the Mainz Bingen Graben, assuming they all were synchronously deposited (Table 3.1). Bibus and Semmel (1977) establish this correlation using the following criteria:

- t_{R4} contains the so-called Mosbach facies,
- t_{R4} contains the same invertebrate fauna as T6,
- normal magnetic polarization of humic clay in the upper parts of t_{R4} and t1 (measurements from outcrop near Werlau for t_{R4} and from t1 in Lower Main Valley, Bibus and Semmel, 1977; see Figure 3.2B),
- pollen spectra of soils in the upper parts of t_{R4} and T6 are characteristic for the younger Cromer interglacial,
- t_{R5} and t2 of the Lower Main Valley show an onset of carbonate-free gravel, and
- t_{R3} and T7 of the Mainz Bingen Graben exhibit a similar gravel content.

This correlation is also favored in recent studies (Fetzer et al., 1995; Hoselmann, 1996), though new data has not been provided since the study of Bibus and Semmel (1977).

Several authors have proposed a morphological correlation:

- T7 equals t_{R5} (Wagner, 1931a; Kandler, 1970),

- Mosbach Sands / T6 equal the uppermost middle terrace t_{R6} (Kandler, 1970), and

- the 195 m terrace of the lower River Nahe valley (younger main terrace) corresponds to t_{R5} and T7 (Andres and Preuss, 1983).

These correlations are based on the observation that T7, t_{R5} and the younger main terrace of the River Nahe all occur at the same elevation of ~ 200 m (Fig. 3.2B). Boenigk (1978; 1987) also supports this correlation, stating that Mosbach Sands/T6 correlate with t_{R5} or t_{R6} in the Upper Middle Rhine Valley and with the upper terraces UT3 or UT4 in the Lower Rhine Embayment because reworked material is frequent in all of these terraces.

3.5. Fault movements

3.5.1. Longitudinal profile

Fault movements can affect the position of fluvial terraces by either horizontal or vertical displacements (strike-slip or extensional faulting) or by a combination of both directions (oblique faulting). None of the previous terrace studies or this study found evidence for strike-slip faulting in the northern URG and adjacent areas. However, as demonstrated in this section, relative elevation differences of terraces have been observed previously in local parts of the study area and have been associated with vertical fault movements (Stäblein, 1968; Semmel, 1969; Kandler, 1970; Semmel, 1978). In order to investigate these local movements and to gain information on the regional trends of vertical motions, a longitudinal profile covering the entire study area from the northern URG to the Upper Middle Rhine Valley was constructed (Fig. 3.6). The profile includes the four terrace groups and the Pliocene surface. The terraces are projected orthogonally onto the present-day longitudinal profile of the Rhine and the lowest and highest morphological position of each terrace group is plotted. The profile follows the entire course of the River Rhine shown in the figures of the study area (Figs. 3.1, 3.2), beginning to the west of Karlsruhe and ending in the Middle Rhine Valley. The morphological correlation is used for the southern areas (Vorderpfalz to WBF scarp) since it gives plausible results for tectonic movements and there is no stratigraphic constraint available. For the area across the HTBF (Mainz Bingen Graben to Upper Middle Rhine Valley) stratigraphic data is available. In Figure 3.6, the stratigraphic correlation has been adopted for this part of the profile because the morphologic correlation does not give plausible results for the reconstruction of tectonic movements. The implications of stratigraphic or morphologic correlation across the HTBF are presented and discussed in detail in section 3.5.2 using Figure 3.7.

General trends

Irrespective of the correlation method adopted across the HTBF, the general trends of terrace elevations in the longitudinal profile are as follows (Fig. 3.6). Relative elevation changes of terrace surfaces are most significant for the older terraces of the profile (uppermost level of middle to higher main terraces), whereas they are minor for the lower and the lowermost middle terraces. This implies that these younger terraces were not affected recently by large-scale faulting or tilting. Considering the older terraces the general trend shows a decrease in elevation in the southern regions (Vorderpfalz to the southern Rheinhessische Hügelland) and a successive increase from the Rheinhessische Hügelland northwards. The highest positions occur in the Upper Middle Rhine Valley. This observation suggests uplift of the northern regions prior to the middle terrace period. Fault movements inferred from the profile are in the following sections separately discussed for each region. The description is supplemented with observations from areas located outside the profile (Mainz Basin and Lower Main Valley).

Vorderpfalz

In the Vorderpfalz, the elevation of terraces decreases systematically in a downstream direction. For all terraces older than the lower terrace, the downstream elevation decrease occurs at a slightly steeper gradient than for the youngest terrace level (Fig. 3.6). This implies that the terraces older than the lower terraces have been slightly uplifted after their formation. The area of strongest uplift is situated in the central part of the Vorderpfalz. The longitudinal profile contains information on the eastward sloping of the higher main and main terraces. This is evident from the large elevation differences between the upper and lower parts of these terraces. Monninger (1985) demonstrated, in W-E profiles along the ridges of the Vorderpfalz, an eastward sloping of the terrace deposits. This sloping is interpreted as resulting from tectonic tilting of the Vorderpfalz. An additional observation is the zigzag pattern of elevations in the southernmost Vorderpfalz. This pattern is related to the morphology of the large ridges that resulted from the erosion of rivers draining from the Pfälzer Wald (Fig. 3.1).

Rheinhessische Hügelland

In the Rheinhessische Hügelland and along the WBF scarp, the elevation of older terraces (older than uppermost middle terrace) increases stepwise at the Pfrimm and the Riederbach Valleys (Fig. 3.6). Vertical displacements of 10 m across the Pfrimm and 20 m across the Riederbach Valley (equal offset for all terraces) are attributed to activity on ~ E-W faults parallel to the valleys (Fig. 3.2A, inset). Fault mapping in the Rheinhessische Hügelland, in particular of E-W oriented faults, is not well constrained and rarely documented (mainly unpublished reports; G. Stahmer, pers. comm. 2005). Leser (1967) proposed a fault parallel to the Pfrimm Valley to explain the pronounced E-W course of its river. Activity on a south-dipping extensional fault along the Seebach Valley influenced deposition of Pliocene sediments (Wagner, 1941b; Heitele, 1971; summary in Franke, 2001), though Pleistocene terraces are apparently unaffected (Fig. 3.6). Results of this study provide evidence for offsets across the Riederbach Valley and proposes significant activity on a previously unknown Riederbach Valley Fault.

WBF scarp

It is evident from the profile of Figure 3.6 that along the WBF scarp the elevation of all terraces, except the lower terraces, increases by a maximum of 30 m from south to north. Land slides have been recognized in the northern part of the scarp (Rogall and Schmitt, 2005), that presumably affected the slopes of the middle and younger main terrace surfaces. For this northern part of the plateau, young tectonic activity has been suggested for the WBF and the faults of the Dexheimer Horst (Sonne, 1969; Sonne, 1972; see inset of Figure 3.2A). In this context, the northern part of the plateau forms a tectonic horst structure. It is therefore likely that young uplift of this horst has caused increased uplift of the terraces to the north of the scarp and lead to the observed slope instability.

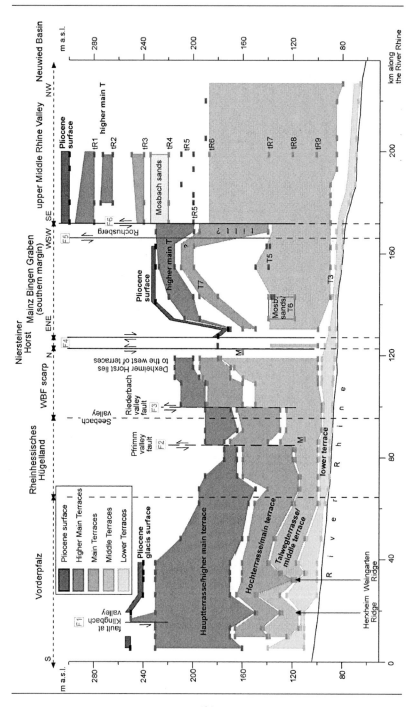

Figure 3.6. Longitudinal profile of terrace groups in the northern URG, Mainz Bingen Graben and Middle Rhine Valley. The lowest and highest morphological position of each terrace group is plotted. The erosional base of the lower terrace, which is partly situated below the present-day riverbed of the River Rhine, is not mapped. Elevations of the lowermost middle terrace in the Upper Middle Rhine Valley are taken from topographic maps. From the low morphological positions of the higher main and main terraces on the Rochusberg it is inferred that tilting due to local tectonics of F5 may have affected the terraces. M = Mosbach Sands, F5 = HTBF, F6 = splay fault of HTBF.

Niersteiner Horst

A major change occurs at the Niersteiner Horst. To the north of the horst, the elevation of the main and higher main terraces is 30 m lower than to the south of the horst (Fig. 3.6). Furthermore, the higher main terrace is 40 m lower on the horst itself than on the WBF scarp to the south. Assuming that mapping of the Mosbach Sands in the south of the horst is correct (outcrop at Oppenheim after Schraft, 1979; marked M in the inset of Figure 3.2A and in Figure 3.6), an elevation difference of 40 m is observed between this locality and the outcrops of Mosbach Sands north of the horst (mapping of Kandler, 1970). Furthermore, outcrops of Oligocene Lower Cerithium beds south of Oppenheim and along the N-S oriented scarp south of Mainz (Laubenheimer Berg) also indicate a 30 m lower position in the north (maps of Steuer, 1911; Sonne, 1989). In summary, all deposits older than the Mosbach Sands occur at a 30 – 40 m higher topographic position to the south of the horst than to the north of it. These observations support the suggestions of Sonne (1969; 1972) that recent tectonic activity was larger on the Dexheimer Horst and the southern shoulder of the Niersteiner Horst than on its northern shoulder. The terrace profile implies that since the time of the higher main terraces the Niersteiner Horst was tectonically inactive.

Selz and Wiesbach terraces in the Mainz Basin

The longitudinal profile shows only terraces of the River Rhine, while the terraces of the rivers Selz and Wiesbach are not included. However, for completeness the Selz and Wiesbach terraces as well as terraces of the River Main are now discussed. The available mapping of Selz and Wiesbach terraces does not cover the entire river courses. The higher main terraces of the Selz are the only terraces that can be traced over longer distances. They show a systematic elevation decrease from the upper to the middle course of the river, which implies that they are unaffected by fault movements

(Fig. 3.2A, inset with heights of terraces). However, in this area, the river courses follow and cross faults bounding the Niersteiner Horst. The linear course of the Selz along the southern fault of the Niersteiner Horst could indicate a fault control on the fluvial incision. This would also support the higher tectonic activity for this fault proposed by Sonne (1969; 1972). However, the present-day mapping of Selz terraces is still insufficient to fully assess possible effects of tectonic movements on its terraces.

Lower River Main Valley

The terraces of the Lower River Main Valley provide information on the vertical component of fault movements (Semmel, 1969; Semmel, 1978). The main and oldest middle terraces (t1 – t3) are frequently offset by extensional faults with vertical offsets of up to 6 m (Semmel, 1969). Displacements of deposits of younger terraces are less frequent and only in the order of few centimeters. On the basis of present-day terrace elevations, Semmel (1978) proposes relative displacements of 40 m of the Cromer age t2 on a segment of the WBF and 20 m of t2 on a splay fault (displacements at Eppstein Horst; Semmel, 1978; F7 in Figure 3.2A).

Mainz Bingen Graben

The profiles in Figure 3.7 display the terrace elevations on both the northern and southern margins of the Mainz Bingen Graben (elevations indicated with numbers in Figure 3.7A, see left part of profiles). The profiles show that the terraces younger than T5 are, ignoring small deviations, at equal elevation along the profile and on each side of the graben. The rise of the base level of the lower terrace from east to west by 4 m after Kandler (1970) has not been included in Figures 3.6 and 3.7 due to the resolution. A significant increase in elevation downstream occurs for the higher main and main terraces and for the Mosbach Sands along the northern side of the graben. On this side, the terraces rise from east to west with approximately equal amounts (Fig. 3.7A, profile on left hand side). The elevation differences between the terraces are in the order of 40 to 50 m. On the southern side, the higher main and main terraces are much closer to each other (20 m difference) and the elevation difference between the younger terraces is much larger than on the northern side (60 m between main terrace and Mosbach Sands; Fig. 3.7A, profile on right hand side). These observations indicate that uplift differed on the sides of the graben. Two uplift phases could explain the observed uplift pattern. 1) The rise of the higher main terraces from 170 to 230 m on the southern side and from 170 to 215 m on the northern side show a difference of 15 m. This indicates that the northern side experienced 15 m more uplift than the southern side after formation of these terraces. Consequently, incision was larger on the northern side and lead to the increased elevation difference between the higher main and main terraces. 2) After formation of the main terraces, the southern side experienced higher uplift. While the rise of the main terraces is about 10 m on the northern side (150 – 160 m), it reaches 50 m on the southern side (150 – 200 m). The N-S differential motion of this second phase would be in the order of 40 m. There are numerous faults both along strike of the Mainz Bingen Graben as well as across the graben. Various authors have mentioned that these have caused changes in terrace elevations, but precise mapping of the faults is not available (Kandler, 1970; Abele, 1977; Sonne, 1977). In this setting, it is therefore likely that the differential uplift of terraces included the contribution of a number of faults. In detail, the uplift was probably much more complex and not as simple as explained here with two phases.

Upper Middle Rhine Valley

Based on terrace mapping in the Upper Middle Rhine no major tectonic displacements affected these terraces. In particular, the elevations of the t_{R4} and t_{R5} terraces are parallel and nearly horizontal between Bingen and the Neuwied Basin (Bibus and Semmel, 1977; Semmel, 1983). For this reason, Bibus and Semmel (1977) suggested an en-bloc uplift of the entire Upper Middle Rhine area (Rhenish Massif).

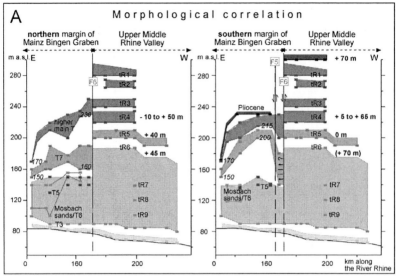

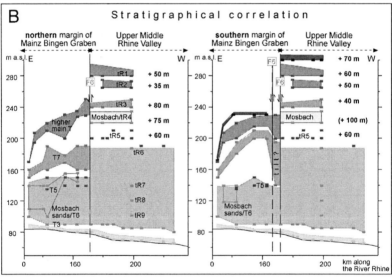

Figure 3.7. A) The concept of the morphological correlation is based on the continuous distribution of main terraces from the southern margin of the Mainz Bingen Graben to the Upper Middle Rhine Valley. The same terraces are displaced by 40 m from the northern margin to the Middle Rhine. Numbers in bold refer to relative uplift of Middle Rhine Terraces at F6, a splay fault of the HTBF. B) The elevation differences of stratigraphically correlated terraces between the southern margin of the Mainz Bingen Graben and the Upper Middle

Rhine have been estimated between the east of F5 and the first outcrop in the Middle Rhine (Assmannshausen of Bibus and Semmel, 1977) in order to avoid the local tectonics of the Rochusberg (see also Figure 3.2B). Between the northern margin and the Middle Rhine, the two outcrops closest to F6 have been used. Numbers in bold refer to relative uplift of Middle Rhine Terraces at F6. Both profiles show offsets between 35 – 80 m.

3.5.2. Fault movements of the HTBF

The most significant fault movements along the longitudinal profile can be inferred for the boundary between the Mainz Bingen Graben and the Upper Middle Rhine Valley that coincides with the HTBF. However, the largest movements do not occur along the HTBF itself (Figs. 3.6, 3.7; F5) but on a parallel fault zone that is presumably related to the HTBF (F6). This fault zone is located at the entrance of the River Rhine to the Upper Middle Rhine Valley (Fig. 3.2B). Depending on the application of morphological or stratigraphic criteria for correlating the terraces across this fault zone, the magnitude of its displacement changes. In order to demonstrate these differences, fault displacements for each correlation scenario have been reconstructed using the terrace data (Fig. 3.7). The following section describes this reconstruction. The implications of these displacement scenarios are discussed thereafter in section 3.6.

Using the morphological criteria for the correlation between the main terraces of the southern margin of the Mainz Bingen Graben (east of F5) and those of the Upper Middle Rhine Valley (west of F6), these terraces occur at the same elevation (Fig. 3.7A, profile to the right; after Wagner, 1931a; Kandler, 1970; Boenigk, 1978; Andres and Preuss, 1983; Boenigk, 1987). However, using the same correlation, there is a 40 m offset for these and the younger terraces (T6 and t_{R6}) between the northern margin and the Upper Middle Rhine Valley (Fig. 3.7A, profile to the left). However, a number of uncertainties with this correlation scheme remain. The correlation of the T6 and t_{R6} terraces between the southern margin and the Middle Rhine cannot be constrained since outcrops of T6 are missing to the west of F5. Additionally, as for the correlation of older terraces, it remains unclear which of the multiple higher main terraces of the Upper Middle Rhine correlates with the single higher main terrace in the Mainz Bingen Graben.

The stratigraphic correlation shown in Figure 3.7B is based on studies by Bibus and Semmel (1977) and is supported by recent reviews of Fetzer et al. (1995) and Hoselmann (1996). From the northern margin of the Mainz Bingen Graben to the

Upper Middle Rhine Valley, offsets vary from 35 – 80 m (Fig. 3.7B, profile to the left). The youngest terraces affected by F6 are the T5 and t_{R5} terraces with 60 m displacement. From the southern margin to the Upper Middle Rhine Valley, offsets decrease from 70 to 40 m until the main terraces (Fig. 3.7B, profile to the right). However, the next younger terraces (T5 and t_{R5}) show an increased offset (60 m). In summary, the latest vertical offset across the HTBF for both sides is in the order of 60 m.

The displacement scenario based on a morphological correlation demonstrates that this correlation scheme contains a number of inconsistencies, especially when comparing individual terraces. However, at a large morphological scale, this correlation provides a good fit. Based on the schematic morphological profile of the terrace staircase shown in Figure 3.3, the middle terraces form a sequence of small surfaces on the slope of the profile while the main terraces build the wide surfaces on the plateau. This morphology is seen both in the Mainz Bingen Graben and the Middle Rhine Valley, with the only difference being that in the Middle Rhine the slope is steeper and the valley is narrower. In particular, the youngest of the plateau terraces in the Middle Rhine Valley, the younger main terrace (t_{R5} of Bibus and Semmel, 1977), is well expressed with large surfaces along the entire course of the valley. It can also be traced along the tributaries of the River Rhine in the area of the Rhenish Massif (rivers Lahn and Mosel). For this reason, Meyer and Stets (1998) used this terrace for geomorphological mapping and inferred uplift rates from its elevation with respect to the recent river bottom. Applying this method to terraces across the HTBF involves using the morphological correlation scheme. Here, the consistent morphology and the equal height of terrace surfaces across the HTBF between the southern Mainz Bingen Graben margin and the Middle Rhine suggest that this fault was inactive since terrace formation and that the entire region experienced an equal amount of fluvial incision. In the Middle Rhine area, incision is thought to be primarily a response to tectonic uplift. In particular, the increased incision during the middle terrace period, which caused formation of the narrow, terraced valley in the Middle Rhine, indicated a major increase in uplift rate during that period (Meyer and Stets, 1998; Van Balen et al., 2000). According to the morphological terrace correlation across the HTBF, the amount of uplift would have been equal in the Middle Rhine and the Mainz Basin, i.e. an en-bloc uplift of the entire region without activity on the HTBF.

Applying the stratigraphic correlation concept, the evolution of uplift and incision in this region would have been different. Petrographic criteria and paleomagnetic dating strengthens the case for a stratigraphic correlation of terraces across the HTBF. According to this concept, the Matuyama/Brunhes boundary at 780 ka lies in the Mainz Basin and the northern URG within the middle terrace period and in the Upper Middle Rhine Valley within the main terrace period (Fig. 3.8). This means that the main and middle terraces in the Mainz Basin and northern URG do not coincide in age with the main and middle terraces in the Upper Middle Rhine Valley (see Table 3.1). Furthermore, this implies for the landscape evolution of the region that increased fluvial incision in response to tectonic uplift occurred first in the Mainz Bingen Graben, leading to formation of the sequence of middle terraces (slightly before 780 ka; Fig. 3.8A). In contrast, the middle terraces of the Upper Middle Rhine were formed later, i.e. after the Matuyama/Brunhes boundary (Fig. 3.8B). Several terrace studies along the Middle Rhine and the River Meuse document a phase of accelerated uplift of the Rhenish Massif just after the Matuyama/Brunhes boundary (Meyer and Stets, 1998; Van Balen et al., 2000). Its uplift of the last 780 ka involved its regional doming with the largest vertical motions of 250 m located in the center of the Rhenish Massif (Meyer and Stets, 1998). Towards the rim of the dome, the uplift decreases to 50 m. In the study area, 100 m are documented for the Mainz Bingen Graben and 50 m for the Lower Main Valley and the central part of the Mainz Basin (Meyer and Stets, 1998). For the relative uplift between the central Rhenish Massif and the Mainz Basin various values are given in literature: 170 m (Wagner, 1931a), 200 m (Meyer and Stets, 1998) and 240 m or more (Sonne and Weiler, 1984). According to Meyer and Stets (1998), the uplift of the Rhenish Massif involved tilting of individual blocks at different rates. This is not in accordance with en-bloc uplift across the HTBF as suggested from morphological terrace correlation. Given the large values of relative uplift that are documented for the study area (170 to 240 m), it is however more likely that the HTBF was reactivated during the uplift of the Rhenish Massif. This is in accordance with the stratigraphical terrace correlation across the HTBF. At this stage, the stratigraphic correlation scenario across the HTBF is preferred and presented in Table 3.1, owing to constraints provided by lithological and paleomagnetic data.

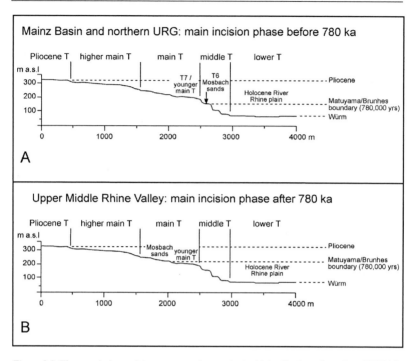

Figure 3.8. The morphology of the terrace staircases in the Mainz Basin and northern URG (A) and the Upper Middle Rhine Valley (B) is the same. However, paleomagnetic dating of the Mosbach Sands / T6 and the t_{R4} main terrace suggest that the main incision phase, which lead to the formation of the steep, terraced slopes, occurred first in the Mainz Basin and northern URG and later in the Middle Rhine.

However, the implied 60 m of Mid-Pleistocene displacement across the HTBF has not left a signature in the morphology. Moreover, the surface traces of the HTBF and related faults (F5 and F6) do not form a fault scarp (see profile in Figure 3.2B). At a large scale, however, a morphological separation between the low level of the Mainz Bingen Graben and the uplifted Rhenish Massif is evident. This separation cannot be attributed to a single fault structure. Thus, the boundary between both regions is a zone of distributed fault displacements. In fact, separate faults can also explain the difference between the northern and southern terrace stratigraphies of the Mainz Bingen Graben, as discussed above.

3.6. Conclusions

The main motivation for this study was to determine the effects of potential fault movements on Pleistocene terraces with the aim of better constraining the characteristics of Pleistocene tectonic activity in the northern URG and the adjacent areas. In order to infer displacement rates from a longitudinal profile it is obvious that absolute ages of terraces are vital. Lack of such data leads to the number of uncertainties in terrace correlation. Keeping these uncertainties in mind, this study uses the available data to provide a first order quantification of tectonic movements. The rates calculated herein rely on the absolute dating of the Mosbach Sands and should be considered as rough estimates. Table 3.2 shows that for the past 800 ka displacements along single faults vary between 10 and 60 m, yielding vertical displacement rates of 0.01 – 0.08 mm/yr, with a tendency for displacement rates to increase from south to north. For a comparison, the subsidence and uplift rates of the *Heidelberger Loch* and the Rhenish Massif are given respectively. The study area, located in between these zones of high subsidence and uplift rates, exhibits displacement rates up to an order of magnitude lower.

Based on the longitudinal profile it can be demonstrated that tectonic activity had the greatest impact on terrace levels during Early to Middle Pleistocene times (Fig. 3.6). Assuming constant tectonic activity at the lowest rate calculated (0.01 mm/yr) displacements accumulated over the last 400 ka would be in the order of 4 m. This is undetectable in a longitudinal profile at the resolution used. Thus, no constraint on low displacement rates can be provided with the method used in this study. Since the youngest terraces (lowermost middle and lower terraces) are for the most part of the longitudinal profile unaffected by faulting, it can be excluded that a significant increase in tectonic activity occurred since Late Pleistocene. Several field studies in the northern URG and the Middle Rhine area confirm a low level of Late Pleistocene tectonic activity. In the Lower Main and Upper Middle Rhine Valleys, displacements of Middle to Late Pleistocene terraces are infrequent and of few centimeters only (Semmel, 1969; Semmel in Illies et al., 1979). Studies in the northern URG show Late Pleistocene rates in the order of 0.03 mm/yr, which is about the mean of the range for the Early to Middle Pleistocene (Vorderpfalz, Monninger, 1985; WBF scarp, *Chapter 2 of this thesis*).

Area	Locality, fault	Youngest terrace/s involved	Age used (ka)	Displacement (m)	Displacement rates (mm/yr)
Rheinhessische Hügelland	Pfrimm valley	uppermost middle terrace	700	10	0.014
	Riederbach valley	uppermost middle terrace	700	20	0.029
WBF scarp	height difference along WBF scarp	younger main	800	20	0.025
	height of WBF scarp	younger main	800	50 - 100m incision	0.063 - 0.125
Niersteiner Horst	height difference between east and west of horst	Mosbach	780	30	0.038
Lower Main Valley (1)	WBF (eastern border of Eppstein Horst) N-S fault	Early Pleistocene gravel	2 Mio (Olduvai – Gilsa event)	200	0.1
	eastern border of Eppstein Horst (NE-SW fault)	t2	400	40	0.1
	western border of Eppstein Horst (NW-SE fault)	t2	400	20	0.05
HTBF	morphological correlation scenario, incision rate	tR5	750	120 m incision, HTBF inactive	0.16
	stratigraphic scenario	tR5	750	60	0.08

Area	Method	Age of deposits/ terraces used (ka)	Displacement rates
Entire study area (summary of regions above)	Terrace mapping (along profile of Fig. 7)	400 – 800	0.01 – 0.08
Northern URG (eastern border, 2)	Thickness of deposits (Heidelberger Loch)	1800	0.2
Rhenish Massif (Middle Rhine, 3)	Incision rates, terrace mapping (values from rim to center of doming)	800	0.1 – 0.3

Table 3.2. Overview of displacement rates calculated from terraces in Figure 3.6 with supplemented data from the Lower Main Valley, the eastern border of the northern URG and the Rhenish Massif. Sources: 1 Semmel (1978); 2 Bartz (1974); 3 Meyer and Stets (1998), Van Balen et al. (2000).

Two locations along the longitudinal profile with up to 10 m of displacements of young terraces need investigation in more detail, in particular coring. In the Mainz Bingen Graben, the base level rise of 4 m of the lower terrace has been interpreted as tectonic in origin though the fault related to the motion is unknown (Kandler, 1970; base level rise not included in Figure 3.6). The second interesting location is the Niersteiner Horst, where the lowermost middle and lower terraces are rising about 10 m (Fig. 3.6). Mapping of these terraces was based on morphological criteria, whereas the elevation of the lowermost middle terrace was mapped at the foot of the slope (compare Figures 3.2A inset and 3.5). In this area of the Niersteiner Horst and the WBF scarp, the east facing slopes with the sequence of middle terraces are relatively steep and slope wash covers the foot of the slope. Thus, it is difficult to distinguish the morphology of the terrace surfaces in this area and overestimating their elevations by a few meters is likely. If recent tectonic activity caused the large displacements observed at these two sites this would imply a significant increase in tectonic activity at order of magnitude higher rates for the Late Pleistocene (0.1 – 0.2 mm/yr).

In summary, this study uses interpretations of topographic maps and local field observations to map terraces in the northwest of the northern URG. The correlation of these terraces with previously mapped terraces to the north and south provides a first order quantification of tectonic movements. The largest uplift of terraces is observed for the terraces bounding the Mainz Basin. The fault structures that contribute to this uplift are the WBF, the Niersteiner Horst and the HTBF. In addition, the displacements of terrace surfaces point to activity of two E-W oriented extensional faults in the Rheinhessische Hügelland to the south of the Mainz Basin. One of the faults, the Riederbach Valley Fault, was previously unknown and its existence is proposed herein for the first time. The motions and displacement rates calculated for the active faults in the study area indicate deformation rates of 0.01 – 0.08 mm/yr. It is suggested that during the Late Pleistocene, deformation occurred at this low rate. The terrace study demonstrates that regional, discontinuous uplift occurred on the western border of the northern URG during the Quaternary, with a significant pulse of uplift in the northern URG during the Early Pleistocene and subsequent during the Middle Pleistocene in the Middle Rhine area (Fig. 3.8).

CHAPTER 4

TECTONIC GEOMORPHOLOGY OF THE NORTHERN UPPER RHINE GRABEN[1]

4.1. Introduction

This chapter addresses the effects of faulting activity on the landscape evolution of the northern Upper Rhine Graben (URG). Currently, the URG and the Lower Rhine Graben are the most active graben segments of the European Cenozoic Rift system. In the URG, Quaternary deposits reach thicknesses from a few tens of meters to over several hundred meters. The main Quaternary depocenter with 380 m thick deposits is the *Heidelberger Loch* situated in the northern URG (Bartz, 1974). In this part of the graben, differences in the thickness of Quaternary deposition were most significant and lead to an asymmetric graben fill with accumulation concentrating mainly adjacent to the Eastern Border Fault (EBF; Bartz, 1974). The present-day surface morphology of the northern URG valley is also asymmetric: The eastern side is a low relief valley while the western side, in particular the Vorderpfalz, is characterized by generally higher elevations (Fig. 4.1). This asymmetry is the result of regional uplift of the western margin and coeval subsidence of the eastern part of the graben, as revealed by geomorphological investigations on fluvial terraces and by the distributions of Quaternary sediments as derived from borehole and reflection seismic data (Stäblein, 1968; Bartz, 1974; Brüning, 1977; Monninger, 1985; Haimberger et al., 2005). Previous studies of the geomorphology of the northern URG area mainly focused on the effects of climatic conditions on landscape evolution during the Quaternary (Stäblein, 1968; Leser, 1969; Brüning, 1977; Klaer, 1977; Barsch and Mäusbacher, 1979; Andres and Preuss, 1983; Höhl et al., 1983; Dambeck and Thiemeyer, 2002).

---

[1] This chapter is partly based on the publication: Peters, G., Van Balen, R.T., 2007. Tectonic geomorphology of the northern Upper Rhine Graben, Germany. Global Planetary Change 58: 310-334.

The contribution of tectonic activity was investigated in a few local studies of marginal areas of the graben (Semmel, 1978; 1979; Beck, 1989; Semmel, 1991; Eitel, 2003). To date, no investigation has been performed on the effects of intra-graben fault activity on the evolution of the landscape and drainage system (River Rhine and tributaries) of the valley.

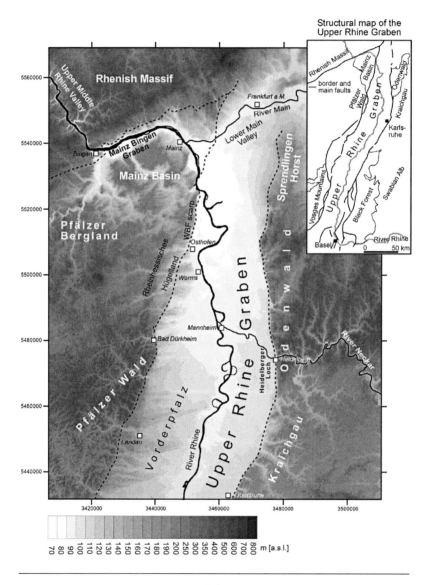

Figure 4.1. Morphology of the study area. The eastern side of the graben in the northern URG is characterized by a low relief with elevations of 85 – 110 m a.s.l. whilst the western side has a landscape of valleys and ridges with elevations of 110 – 250 m a.s.l. The shoulder areas reach peak elevations of greater than 600 m a.s.l. in the Pfälzer Wald, the Odenwald and the Rhenish Massif. The Kraichgau, Sprendlingen Horst and the Mainz Basin are lower shoulder areas with elevations generally between 200 – 300 m high. The structural map displays the border faults of the URG and the main structural units in the shoulder areas of the graben. The course of the River Rhine across the URG is indicated with the grey line. Digital data for the contour map and shaded relief maps in Figures 4.1, 4.2, 4.6 and 4.11 from DGM50 M745 (2001). All maps are in German Gauss-Krüger coordinate system.

The impact of crustal deformation on a landscape is expressed in the form of various tectono-geomorphic features: escarpments, discontinuities along river courses, changes in drainage patterns, channel types, flood plain morphology, deformation of river terraces and the morphology of active mountain fronts (e.g. Ouchi, 1985; Keller and Pinter, 1999; Schumm et al., 2000; Burbank and Anderson, 2001). The intensity of the interplay between tectonic and surface processes that form a tectonic geomorphology is strongly controlled by climatic conditions (Bull, 1991). During the Quaternary, the level of tectonic activity in the northern URG area was relatively low with subsidence rates reaching a maximum of 0.2 mm/yr in the *Heidelberger Loch*, but being approximately an order of magnitude lower in the rest of the graben. In addition, the region is characterized by multiple Pleistocene climate changes (glacial and interglacial periods) and recent, intensive human modifications. Given this setting, the records of Quaternary climate-driven surface processes and human activity are presumably better preserved than the record of the relatively slow tectonic deformation. This further implies that the effects of Quaternary upper crustal deformation on the landscape, typically generated by faulting, tilting, folding or relative uplift, are rather subtle in the northern URG. A simple calculation helps to highlight the scale of expected tectonic records. Considering an average deformation rate for the northern URG of 0.03 mm/yr (see *Chapter 3*), 1 m of continuous uplift will require 30 ka. This relief is likely being eroded during such a large time span.

This chapter focuses on the effects of faulting activity on the morphology and drainage system of the northern URG. The aim of the study is to detect active faults and to specify which aspects of the geomorphology of the northern URG are of tectonic origin and which are related to fluvial processes. The area investigated is much larger than those of previous studies. It includes the Rhine Valley and the graben shoulders

(Kraichgau, Odenwald, Pfälzer Wald, Mainz Basin and southern Rhenish Massif; Fig. 4.1). This chapter comprises three parts. The first part is a review of the drainage system evolution, landscape formation and tectonic processes beginning with the onset of a fluvial environment in the northern URG during the Late Miocene and ending with the present-day situation. The evolution is presented as a series of paleogeographic maps based on a data compilation from previous studies and new terrace mapping, sediment distributions and fault mapping. This part of the study gives an overview of long-term tectonic and fluvial processes. The second part investigates the effects of the structural setting on the present-day landscape. The overlay of topographic features, drainage patterns and faults is used to investigate regional scale relationships between drainage and fault orientations and between morphology and faults. The third part uses quantitative measurements of the landscape shape in order to determine the balance between erosion and tectonic activity in the area. This method includes calculation of geomorphic indices from the present-day morphology and drainage system, which mark changes in valley shape, mountain-front shape and in slope gradient of river profiles. These indices are one method for characterizing lithological or tectonic controls on the morphology and drainage and enable determination of the balance between erosion and tectonic activity. Furthermore, this method allows for a distinction between relatively active and inactive segments of the border faults and to detect local deformation on river profiles. The results are discussed and integrated in the last section of this chapter.

4.2. Regional setting

4.2.1. Present-day drainage system

The present-day River Rhine flows along the central axis of the URG (Fig. 4.1). In the southern URG, between Basel and Strasbourg, the river has a braided pattern in a wide floodplain. At the entrance of the River Rhine into the southern URG a reduction in stream gradient causes deposition of its bed load (gravel and boulders) and the development of a braided river pattern. The northward slope of the URG valley decreases gradually from south to north (0.08 to 0.03%). In the northern URG (north of Karlsruhe), the slope is low enough for the River Rhine to develop a meandering pattern. The sediment load of this part of the Rhine consists of gravel and sand. In the floodplain meadows and oxbow lakes, clay and silt deposition is characteristic (e.g.

Pottgiesser and Halle, 2004).

The flow of the River Rhine is NNE along the majority of the URG (Fig. 4.1). A change in flow direction to the NNW occurs in the northern URG near the city of Mannheim (Fig. 4.2). From this location towards the northern end of the URG, the valley narrows and the River Rhine occupies the western side of the valley. A large deviation from this NNW flow occurs near the town of Eich, where the river takes a wide bend to the east (Eicher Bogen). At the city of Mainz, after confluence with the River Main, the Rhine changes its course to a WSW-ENE direction until Bingen. Here, the river enters the Rhenish Massif and continues its course in a NW direction. After the Rhenish Massif, the River Rhine enters the Lower Rhine Graben and finally the North Sea.

Important tributaries of the River Rhine in the northern URG include the rivers Neckar and Main (Fig. 4.2). Closer to the Rhenish Massif, the rivers Selz and Nahe, with catchment areas in the Mainz Basin and Pfälzer Bergland respectively, flow into the River Rhine. A large number of smaller tributaries drain the graben shoulders and the sides of the northern URG valley. These smaller streams show consistent changes from a dendritic pattern in the shoulder areas to a parallel pattern within the URG (Fig. 4.2). The parallel pattern is particularly well developed by a high density of ENE–trending streams in the Vorderpfalz. This contrast in the drainage network is related to the considerable topographic change from relatively steep topographic gradients in the shoulder to a low gradient alluvial landscape in the graben.

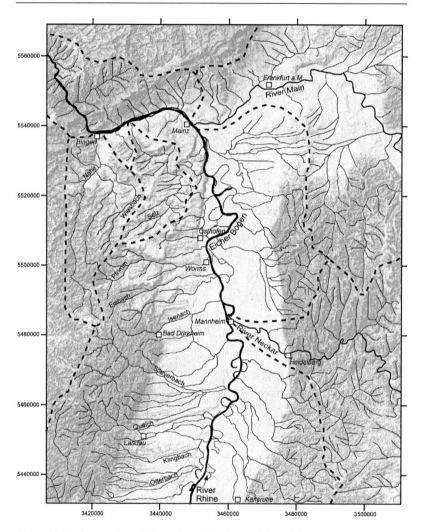

Figure 4.2. Drainage system of the northern URG. Water divides of the catchment areas of the rivers Rhine, Neckar, Main, Nahe and Selz are indicated with dashed lines.

4.2.2. Geological setting

Figure 4.3 displays the geology of the northern URG and shoulder areas. The oldest rocks in the area are crystalline basement rocks, which are exposed in the Odenwald (Fig. 4.3A). The sedimentary cover is of Permian to Jurassic age and consists mainly of sandstones, marls and limestones. Siliciclastic formations are exposed on the western graben shoulder in the Pfälzer Wald and Pfälzer Bergland and on the eastern shoulder to the south, east and north of the Odenwald. Its northern part with Rotliegend sediments is referred to as the Sprendlingen Horst. Carbonates are exposed in the Kraichgau in the southeastern part of the study area. Tertiary deposits to the NW and NE of the URG are related to sub-basins, which are presently located outside of the graben. During the early rifting stages (Eocene to Oligocene), the Mainz Basin formed part of the URG. A distinct Mainz Basin developed during the Late Oligocene, and subsequently evolved independently from the URG, in so far as it was involved in the episodic uplift of the Rhenish Massif (Meyer et al., 1983). This relative uplift brought the Mainz Basin in a marginal location of the URG. The Cenozoic graben fill in the northern URG and the Mainz Basin consists of a sequence of fluvial and lacustrine sediments, interrupted by marine sediments. Since the Early Miocene, the depositional environment was fluvial and lacustrine. The thickness of the intra-graben sediments reaches a maximum of 3200 m near Mannheim in the central part of the northern URG, whereas towards the margins of the graben and in the Mainz Basin thicknesses reduce to a few hundred meters (Doebl and Olbrecht, 1974; Fig. 1.3). During the Quaternary, accumulation of mainly fluvial sediments continued in the northern URG. The largest deposition is documented in the *Heidelberger Loch* located along the Eastern Border Fault (> 380 m, Bartz, 1974). Towards the western and northern sides of the graben, the thickness of Quaternary deposits decreases to 20 m. The Rhenish Massif to the NW of the URG consists of Devonian and Carboniferous sediments that were strongly folded during Variscan orogeny (Fig. 4.3A). In this chapter, only the southern part of the Rhenish Massif is included in the area of investigation. Here, the rocks comprise mostly low-grade metamorphic shales (Hunsrückschiefer; Meyer and Stets, 1996). Near the southern border of the massif, quarzites are found (Taunusquarzit). Phyllites of possibly Siluarian age exist along the Hunsrück-Taunus Boundary Fault (HTBF, Anderle, 1987; Fig. 4.3A). The HTBF is a Variscan terrane boundary that has been reactivated during URG formation (Anderle, 1987; Rheno-Hercynian/Saxo-Thuringian suture in Ziegler and Dèzes, 2005). Associated with the HTBF is the so-called Mainz Bingen Graben (Wagner, 1930), which is a small graben structure, located along strike of the HTBF between Mainz and Bingen (Fig. 4.3A). A distinct tectonic feature within the Mainz Basin is the Niersteiner Horst. Tertiary to Quaternary uplift of the horst has exposed the Permian sediments now forming the horst structure (Sonne, 1969). Recent tectonic activity is concentrated on the southern boundary fault of the Niersteiner Horst, which joins the WBF north of the River Rhine (Sonne, 1969; 1972).

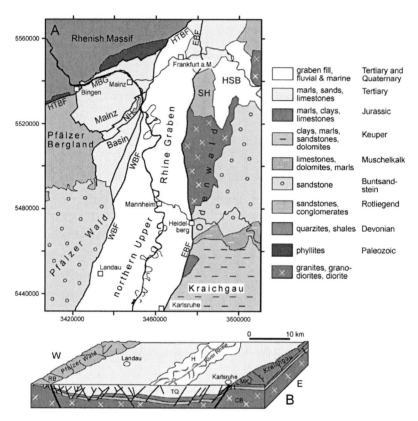

Figure 4.3. A) Geological map of the northern URG and adjacent shoulder areas. EBF – Eastern Border Fault, HSB – Hanau-Seligenstadt Basin, HTBF – Hunsrück-Taunus Boundary Fault, MBG – Mainz Bingen Graben, NH – Niersteiner Horst, SH – Sprendlingen Horst, WBF – Western Border Fault. Geology after GÜK300 (1989) and GÜK500 (1998). B) Simplified geological profile across the northern URG (modified after Illies, 1974b). H – Holocene, TQ – Triassic, Quaternary, MKJ – Muschelkalk, Keuper, Jurassic, RB – Rotliegend, Buntsandstein, CB – crystalline basement

4.2.3. Fault system

The northern URG is a highly faulted area (Fig. 4.4). Geological mapping of the surface and shallow subsurface has identified the faults of the shoulder areas (e.g. Illies,

1974a; Tietze et al., 1979; Stapf, 1988). Intra-graben fault mapping is based mainly on interpretation of seismic sections collected during a period of intensive hydrocarbon exploration of the ~ 1 – 2 km deep Oligocene and Miocene reservoirs during the 1960s and 1970s (fault data in Figure 4.4 compiled after Andres and Schad, 1959; Straub, 1962; Behnke et al., 1967; Breyer and Dohr, 1967; Illies, 1967; 1974a; Derer et al., 2003). Assuming 60° fault dips, surface traces of the mapped faults would lie within ~ 1 km of the faults mapped at depth (Fig. 4.4). Unfortunately, little emphasis was placed on mapping the faults in the post target interval and current knowledge of the faults within Pliocene and Quaternary units of the northern URG is poor. Recent high-resolution reflection seismic surveys on the rivers Rhine, Main and Neckar focused on the structural architecture of Pliocene and younger deposits and provide new insights into younger intra-graben fault activity (Haimberger et al., 2005; G. Wirsing, pers. comm. 2005). Faults with documented Quaternary activity are highlighted in Fig 4.4.

Significant differences can be observed between the intra-graben and shoulder fault pattern. Intra-graben faults strike dominantly NNW with subsidiary NNE and NW sets (Fig. 4.4, inset A). Fault strikes in the shoulder areas are more diverse, with orientations between NW and NE having a population maxima NNE representing the border faults and adjacent parallel fault strands (Fig. 4.4, inset B). The kinematics of intra-graben faults is thought to be dominantly extensional (Fig. 4.3B) although several authors argue for a total of few kilometers of horizontal displacement on them (e.g. Bosum and Ullrich, 1970; Meier and Eisbacher, 1991; Ziegler, 1992; Groshong, 1996; Laubscher, 2001). It is important to note that mainly 2D seismic data is available and thus any determination of the horizontal displacement is problematic. The total vertical displacement of intra-graben faults of several hundred to thousands of meters is much better constrained (e.g. Pflug, 1982). Displacements in Quaternary deposits on intra-graben faults have been reported to be in the order of tens of meters (Haimberger et al., 2005; G. Wirsing, pers. comm. 2005).

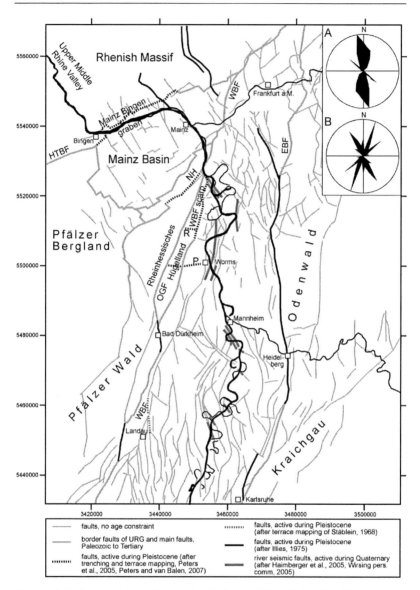

Figure 4.4. Fault map of the study area. R – Riederbach Valley Fault (after Chapter 3), P – Pfrimm Valley Fault (after Leser, 1967; Chapter 3). Roseplot A shows orientations of all intragraben faults, N = 156. Roseplot B shows orientations of border faults and all faults outside the graben, N = 94. Orientations were calculated on fault segments using "rose tool" of the software ArcView .EBF – Eastern Border Fault, HTBF – Hunsrück-Taunus Boundary Fault, NH – Niersteiner Horst, OGF – Oppenheim-Grünstadt Fault, WBF – Western Border Fault.

4.2.4. Summary of the Rhine system since Late Miocene in context of northern Upper Rhine Graben tectonics

This section summarizes the evolution of the Rhine system in the study area in context of the tectonic processes. The data presented here as a series of paleogeographic maps is compiled from previous geomorphological and sedimentological studies (Fig. 4.5). It is supplemented with new, fault and terrace mapping data, all of which is combined to yield revised and/or new interpretations. In this thesis, Pleistocene fluvial terraces in the northern URG, Mainz Basin and Middle Rhine Valley have been investigated in order to infer Pleistocene tectonic activity from displaced terrace surfaces (*Chapter 3*). After a review of existing terrace studies, a regional correlation of the terraces was presented in *Chapter 3* and a simplified nomenclature of four terrace groups was proposed (higher main, main, middle and lower terraces), that is adopted in this chapter. New mapping of terraces along the eastern margin of the Mainz Basin (WBF scarp and Rheinhessische Hügelland, Figs. 4.4, 4.5) is also included. Age determination of most terraces in the area is relative and based on the elevation of terrace surfaces. Ages were assigned by correlation with Quaternary glacial periods. Only for a unique middle terrace element, paleomagnetic and paleontological data are available (Kandler, 1970; Bibus and Semmel, 1977). The ages used in this study follow Kandler (1970), Scheer (1978), Grimm (2002) and Boenigk and Frechen (2006). New fault mapping in the Rheinhessische Hügelland is based on displaced terraces across the Pfrimm and Riederbach Valley Faults (Fig. 4.4). Furthermore, paleoseismological investigations (*Chapter 2*) focused on Pleistocene tectonics along the Western Border Fault (WBF). The results of this study are incorporated into the summary presented herein.

Late Miocene, Figure 4.5-1

The River Rhine has flown northwards towards the North Sea since the Miocene. The earliest evidence of River Rhine sediments in the study area are the Upper Miocene Dinotherium sands (Tortonian age, Grimm, 2002). Remnants of these sands show that a paleo River Rhine crossed the Mainz Basin as a wide braided-river system (e.g. Bartz, 1936; Leser, 1969; Brüning, 1977). Paleo-geographical reconstructions propose that the river system extended over the entire northern URG valley (Hagedorn, 2004). From Late Miocene onwards, the Rivers Selz, Wiesbach, Nahe and Pfrimm are tributaries of the Rhine (e.g. Leser, 1969).

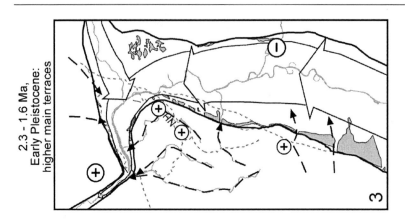

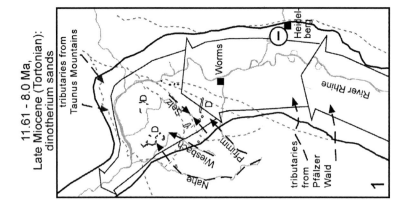

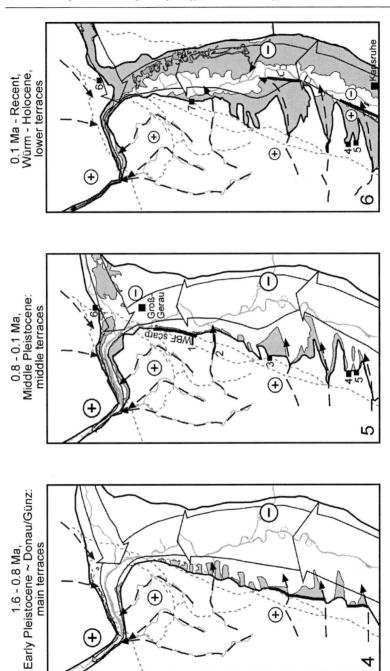

Figure 4.5. Paleogeomorphic evolution of the northern URG in six time steps from Late Miocene to present. Grey units: characteristic fluvial deposits of time steps. Wide arrows: flow directions of the rivers Rhine and Main. Black solid lines: extents of fluvial system. Thick black solids lines: faults active at time steps. Grey dashed lines: major faults. AV-G arvernensis gravel, D – Dinotherium sands, MBG – Mainz Bingen Graben, NH – Niersteiner Horst. Localities with documented tectonic activity during steps 5 and 6: 1 – Riederbach Valley Fault, 2 – Pfrimm Valley Fault, 3 – Forst, 4 – Klingen, 5 – Barbelroth, 6 – Eppstein Horst, 7 – trench site near Osthofen.

Early – Late Pliocene, Figure 4.5-2

During the Early and Late Pliocene, the River Rhine shifted northeastward away from the Mainz Basin due to uplift of the southern Mainz Basin and ongoing subsidence of the Mainz Bingen Graben (Bartz, 1936; 1950; Wagner, 1962; Abele, 1977; Brüning, 1977; Klaer, 1977). From the Late Pliocene onward the principal areas of subsidence are located near Heidelberg and Worms (Weiler, 1952; Wagner, 1962; Bartz, 1974). Clay-rich sands in the Worms area and south of the River Main represent lake environments within the Rhine valley (Weiler, 1952; Wagner, 1962). Relative uplift in the southern Mainz Basin caused reduced accumulation of deposits, karst phenomena and allowed formation of red bed soil horizons (Abele, 1977). At that time, the northern Rhine outlet valley (Middle Rhine Valley) was a wide, shallow valley, the gravel deposits of which now form the oldest terrace preserved in this steadily uplifting region (Kieseloolith-Terrasse of Jungbluth, 1918). Since Early Pliocene, the River Main is present in the northern URG, represented by so-called "arvernensis" gravel (Zanclian age, Grimm, 2002).

Early Pleistocene, Figure 4.5-3

During the Early Pleistocene, fluvial conditions dominated the entire Rhine Valley. Along the western margin of the URG, the deposits are preserved as fluvial terraces, now classified as the higher main terraces (Stäblein, 1968; Kandler, 1970); *Chapter 3 of this thesis*). The presence of terraces on the western margin and the occurrence of polymict Pfälzer Wald material in the center of the URG (Hagedorn, 2004) show that sediment supply from the western shoulder increased in response to its uplift. At the eastern graben margin, fluvial deposition concentrated in the *Heidelberger Loch* depocenter. The River Rhine possibly flowed close to this subsiding part of the graben (Hagedorn, 2004). North of the depocenter, small terrace fragments adjacent to the Eastern Border Fault reflect relative uplift of local areas along the eastern graben margin. In the northerly Mainz Basin, relatively rapid uplift of the Niersteiner Horst caused a division into an eastern and western plateau (Brüning, 1977) and migration of the River Selz northwestwards (Klaer, 1977). This lead to increased incision of the rivers of the Mainz Basin and the establishment of a tableland landscape (Brüning, 1977). Pleistocene terraces of the Middle Rhine occupy a much narrower zone than the Pliocene terraces and span a considerable elevation range (~ 100 m; e.g. Boenigk and Frechen, 2006) suggesting accelerating uplift of the Rhenish Massif.

Early – Middle Pleistocene, Figure 4.5-4

Early – Middle Pleistocene terraces (main terraces) are preserved along the western margin of the northern URG, on the flanks of the Mainz Bingen Graben and in the Middle Rhine Valley (Stäblein, 1968; Kandler, 1970; Bibus and Semmel, 1977; *Chapter 3 of this thesis*). The locations of two faults coincide with terrace scarps in the Vorderpfalz, suggesting a clear influence of fault activity on terrace formation (Stäblein, 1968). The preservation of narrow patches of fluvial deposits as a terrace flight next to the modern River Rhine channel in the Pfälzer Wald, the Mainz Basin and the Middle Rhine is attributed to the interplay between prolonged uplift in these areas and river incision.

Middle Pleistocene, Figure 4.5-5

After the development of the main terraces, a phase of increased incision occurred and a sequence of the middle terraces formed in the northern URG and Mainz Bingen Graben. These terraces are distinct from the main terraces with morphologies characterized by narrower zone, smaller preserved patches and larger elevation differences between levels (Kandler, 1970). In the Middle Rhine Valley, this contrast between the main and middle terraces is even more pronounced (e.g. Boenigk and Frechen, 2006). The increased fluvial incision in the Mainz Bingen Graben and Middle Rhine Valley are interpreted as the response to tectonic uplift (Meyer and Stets, 1998; Van Balen et al., 2000). Based on a review of previous terrace correlation between the URG and the Middle Rhine, it is suggested in this thesis (*Chapter 3*) that tectonic uplift occurred first in the northern URG and Mainz Basin (slightly before 800 ka, based on paleomagnetic data; (Bibus and Semmel, 1977; Boenigk and Frechen, 2006) and shortly after 800 ka in the Upper Middle Rhine Valley. This implies that the transition from main to middle terrace formation is slightly older in the northern URG and Mainz Basin than in the Middle Rhine. During the middle terrace period (~ 0.8 to 0.1 Ma), localized subsidence occurred in the Groß-Gerau area. Subsequently, the River Main shifted southward and left wide middle terraces in this area (Scheer, 1978). In the eastern Mainz Basin, middle terraces are aligned along strike of a segment of the WBF and form a 50 – 100 m high scarp (*Chapter 3*). The alignment of terraces with the WBF, in combination with the morphological expression of the latter is interpreted to reflect coeval fluvial activity and WBF tectonic reactivation (*Chapter 3*). Terrace mapping in the Mainz Basin shows that displacements in the order of tens of meters on main and uppermost middle terraces (post ~ 0.8 – 0.6 Ma) have occurred on the Pfrimm and Riederbach Valley Faults and on faults bounding the southern part of the Niersteiner Horst (*Chapter 3*). Post Mindel faulting (post ~ 0.4 Ma) is recognized in the Lower Main Valley (6 m displacement; Semmel, 1969) and on the HTBF (60 m displacement; (*Chapter 3*). Centimeter-scale displacements after the Riss glacial are documented in the Lower Main Valley (Semmel, 1969), in the Vorderpfalz (decimeter-scale; (Monninger, 1985) and for the WBF bounding the Pfälzer Wald, where loess deposits of Riss age are displaced by 0.5 m (Weidenfeller and Zöller, 1995). In summary, deformation rates inferred from displaced terraces of Early to Middle Pleistocene age range from 0.01 – 0.08 mm/yr (*Chapter 3*).

Würm – Holocene, Figure 4.5-6

The lower terrace, preserved across much of the northern URG, Mainz Bingen Graben and Middle Rhine Valley dating from the Würm glacial, underwent incision by a narrowed River Rhine which continued into the Holocene (Figs. 3.2A, 4.5-6). In the south of the study area, a scarp between the lower terrace surface and the Holocene floodplain is well developed on either side of the Rhine (10 m high Hochgestade scarp near Karlsruhe; Scheer, 1978). Increased fluvial incision into the lower terrace on the western side is interpreted to be associated with young vertical motions along nearby faults (Stäblein, 1968). The western scarp is 4 m higher than the eastern scarp. The Late glacial River Neckar (Bergstrassen-Neckar) abandoned its course, presumably at the end of the Younger Dryas (Dambeck and Bos, 2002), since when its confluence with the Rhine has been located near Mannheim. The Holocene River Rhine flows along the central axis of the graben. The eastern border is no longer the down-tilted side of the graben and the *Heidelberger Loch* may no longer be locus of maximum subsidence.

Indications of Late Pleistocene to Holocene faulting are scarce in the study area and displacements do not exceed 0.5 m. Displacements of Würm loess have been identified in outcrops in the Vorderpfalz (Monninger, 1985), in the Lower Main Valley (Semmel, 1969) and at the trench site of the paleoseismological studies of this thesis (*Chapter 2*). Using this data for calculating deformation rates reveals an average rate of 0.03 mm/yr since Late Pleistocene. The occurrence of rapids on the River Rhine near the Niersteiner Horst and the HTBF is commonly interpreted as an indicator of ongoing tectonic activity on theses structures (Wagner, 1930; 1931a; 1962; Glatthaar, 1976; Rothausen and Sonne, 1984).

4.3. Investigation of tectonic influence on morphology and drainage

Rivers are sensitive to changes in tectonic deformation, adjusting over periods of decades to centuries. Therefore, the drainage system of a region may record the evolution of tectonic deformation (e.g. Ouchi, 1985; Schumm, 1986; Holbrook and Schumm, 1999; Keller and Pinter, 1999). The imprint of tectonic activity on the morphology and the drainage system of the northern URG is now investigated in order to determine the location of fault scarps and tectonically influenced parts of the drainage system. This objective is achieved by conducting an integrated analysis of topography, fault distributions, morphological features and the drainage network.

4.3.1. Morphological expression of faults

Figure 4.6A shows the fault network of the study area including both faults of pre-Quaternary and Quaternary age (compare with Figure 4.4). Fault traces highlighted in Figure 4.6A coincide with scarps or aligned valleys. A rose diagram of fault strike from the northern URG shows that the majority of faults with a morphological expression are oriented NNE (020°; Fig. 4.6A). This is the so-called Rhenish direction. Many of the highlighted fault segments have been previously identified as active during the Quaternary (see Figure 4.4). The most obvious fault related morphological features of the northern URG are the graben borders. On both graben borders, the morphological scarps are at the same location as the surface traces of the border faults. Stäblein (1968) and Zienert (1989) interpret from this observation that tectonic activity of the graben borders along the Pfälzer Wald and Odenwald respectively is at balance with erosional activity.

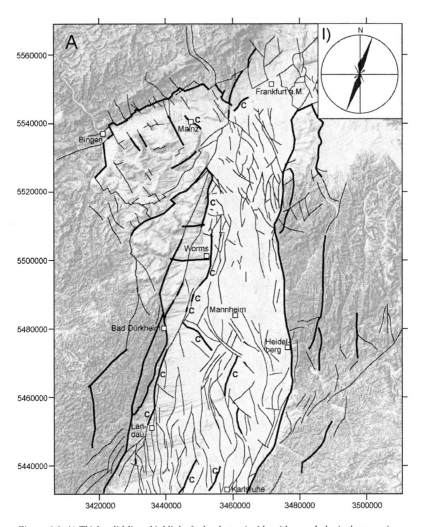

Figure 4.6. A) Thick solid lines highlight faults that coincide with morphological expressions, such as valleys or scarps. Composite scarps of both fluvial and tectonic origin are suggested for the fault segments marked C. The orientations of the highlighted faults are shown in roseplot I). Orientations were calculated on N = 36 fault segments.

Several faults, particularly those associated with the western margin of the graben, coincide with scarps of fluvial terraces (compare the maps of Figures 4.5-3 to

4.5-6 and Figure 4.6A). These "composite scarps" are mapped by integrating existing terrace and sub-surface fault maps. In the study area scarps of combined fluvial dissection and tectonic displacement have only been proposed for a few locations: the WBF scarp (*Chapter 3*), two NNE-trending scarps near Landau (Stäblein, 1968; *Chapter 3*) and the NW-trending scarp of the lower terrace southwest of Mannheim (Frankenthaler terrace scarp; Weidenfeller and Kärcher, 2004; Fig. 4.6A). The results presented here suggest that there area many composite scarps (fault segments marked C on Figure 4.6A).

4.3.2. Structural control on drainage system

Faults that may have influenced the orientation and alignment of river courses are highlighted in Figure 4.6B. Most of these faults and the associated streams have directions between NW-SE and NNW-SSE and are located in the eastern part of the graben and in the southern Rhenish Massif. Structural influence on small streams in the southeastern graben is observed in young (Late glacial and Holocene) parts of the valley, and needs further investigation. South of Mannheim, NW-SE oriented active faults have been identified in recent river seismic profiles (G. Wirsing pers. comm., 2005) and indicate that recent faulting may have imposed controls on NW-SE drainage. Note that these faults have no significant morphological expression (Fig. 4.6A). Further north, the late glacial River Neckar (Bergstrassen-Neckar) appears bounded by faults (as inferred by Wagner 1962, who did not identify these faults). In the Mainz Basin, several large streams follow known faults, which has been previously recognized by Wagner (1930).

The majority of known faults and streams in the western part of the graben cross cut each other, implying minimal structural control on drainage systems of this area. Further examples of minimal structural control occur in the Otterbach, Klingbach and Queich valleys in the Vorderpfalz. These rivers form wide valleys with significant, ENE-WSW trending, linear scarps on the valleys' northern sides (arrows in Figure 4.6B). The linear trend of these scarps suggests a tectonic control. However, comparison of faults, drainage directions and morphological features shows that these linear scarps do not coincide with either known faults or dominant fault trends (Figs. 4.6A, B). This leads to conclude that these scarps are entirely of an erosional nature.

Near the city of Worms along the western side of the graben, two E-W trending faults are proposed, which influenced river courses. Results of recent morphological terrace mapping demonstrate significant vertical displacements across the valleys of the Pfrimm and Riederbach Rivers (*Chapter 3*). These displacements are attributed to activity on the Pfrimm and, previously unidentified, Riederbach Valley Faults. The Pfrimm Valley Fault was postulated by Leser (1967) on the basis of the linear E-W course of the river.

There may be a structural control on the modern River Rhine in the southern part of the study area. Here, the youngest lower terrace scarps on either side of the river coincide with NNE striking faults, suggesting that they influenced the lateral extents of the modern river valley. The origin of the anomalously large Rhine bend at Eich (Eicher Bogen) remains unclear. This bend does not run parallel to mapped faults, but it is situated in an inter-basin transfer zone that subdivides the northern URG in two halfgrabens (a northern and southern sub-basin; Derer et al., 2003). On the base of positive Bouguer gravity anomalies an elevated basement (Stockstadt High/Stockstadter Schwelle after Cloos, 1937) below a relatively thin graben filling has been initially proposed for this region (e.g. Cloos, 1937; Scharpff, 1977). Recent interpretation of seismic and borehole data from the graben filling revealed that the transfer zone represented a structural high and an elevated paleotopographic barrier during the Cenozoic graben evolution. This caused a reduced deposition of Tertiary sediments and particularly in its western part, adjacent to the WBF scarp, a reduced deposition of Quaternary sediments (Derer et al., 2003; Haimberger et al., 2005). In this context, one explanation for the origin of the bend maybe that recently the transfer zone acted renewed as an elevated barrier and caused the Holocene River Rhine to take a wide bend to the east away from the relatively elevated area.

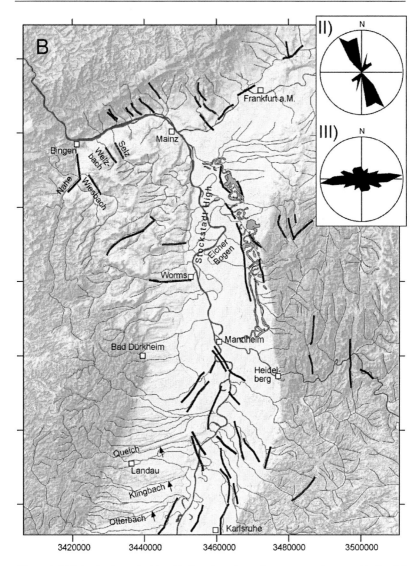

Figure 4.6. B) Faults that coincide with the orientations of streams are shown as black solid lines. Most of these faults have orientations between NW-SE and NNW-SSE as demonstrated in roseplot II) with N = 52 fault segments. The orientations of the entire drainage system are shown in roseplot III) with N = 412 stream segments measured. Arrows indicate scarps at the northern sides of the Otterbach, Klingbach and Queich valleys, which do not coincide with faults in the sub-surface and are thus interpreted as erosional scarps.

4.4. Quantitative measurements of geomorphic indices

4.4.1. Method

Quantitative measurements of a number of geomorphic indices are commonly used as a reconnaissance tool in tectonic geomorphology studies to identify areas experiencing tectonic deformation (Bull and McFadden, 1977; Keller and Pinter, 1999; Burbank and Anderson, 2001). The study of geomorphic indices aims to quantify the present-day landscape shape and to characterize the balance between erosion and tectonics of this landscape. Since rivers are sensitive to tectonic deformation and resulting changes in slope and base level, the stream length-gradient index and the valley shape index are calculated on river profiles and valleys to detect such changes. Additionally, the shape of mountain fronts is described by the mountain front sinuosity. In order to estimate relative variations of tectonic activity in a study area, the combination of different geomorphic indices is used (for a detailed description of the method see Keller and Pinter, 1999; Burbank and Anderson, 2001). Typically, geomorphic indices have been applied to areas of high tectonic activity (e.g. California, Bull and McFadden, 1977; Merrits and Vincent, 1989; Azor et al., 2002, see summary of case studies from California in Keller and Pinter, 1999; Taiwan, Chen et al., 2003; Costa Rica, Wells et al., 1988). In areas of low to medium tectonic deformation, the effects of tectonic processes on the landscape are subtle, but nevertheless can be identified by means of geomorphic indices calculations (Marple and Talwani, 1993; Silva et al., 2003). Recently, tectonic effects on the drainage system of the southern URG have been demonstrated. These studies focused on the analysis of drainage orientations and geomorphic properties of the river network and basin geometries (e.g. stream length-gradient index, drainage density, basin area and slope; Giamboni et al., 2004; Giamboni et al., 2005; Fraefel et al., 2006).

In this study, the calculation of geomorphic indices focuses on the border faults and adjacent shoulder areas of the URG with the aim to assess the degree of tectonic activity on the borders faults. Since the URG is a low activity area, it is expected that the effects of tectonic activity have left only subtle imprints on the present-day landscape. The following indices were calculated:

- Slope gradient index to determine changes of slope gradients of streams crossing both the border faults and intra-graben faults,
- Mountain-front sinuosity index to characterize the morphology of the mountain

fronts along the border faults, and

- Valley shape index of valleys in the shoulder areas adjacent to the border faults to evaluate the level of fluvial incision as a response to uplift.

The quantitative measurements of the indices were performed using digital elevation data. In this study, a digital elevation model (DEM) DGM50 M745 with 30 m horizontal and 3 – 5 m vertical resolution was used (DGM50, 2001). For measurements on the WBF scarp, a DEM with 20 m horizontal and 0.5 m vertical resolution was used (DGM5, 2002).

4.4.1.1. Indices of stream gradient changes

According to the "graded streams" theory of Mackin (1948), streams that are in equilibrium with uplift, erosion and deposition reach stable concave longitudinal profiles with a downstream decreasing slope. Deviations from such a graded profile can result from multiple factors such as erosion resistance of rocks, shifts in sediment supply and tectonic activity. In order to attribute gradient changes of river profiles entirely to tectonics the other factors have to be eliminated as causes (e.g. Holbrook and Schumm, 1999). Gradient changes at a particular reach of the river are proportional to changes in stream power. An increase in gradient and thus steepening of the slope can amongst others be caused by an increase in stream power due to juncture with a major tributary stream. In order to highlight gradient changes Hack (1973) developed the stream length-gradient index (SL). The SL index is defined as the product of channel slope at a given point and the channel length to the source of the stream:

$$SL = \frac{\Delta H}{\Delta L} \times L$$
                                                         Eq. 4.1

where $\Delta H/ \Delta L$ is the gradient of the stream segment investigated and L is the total upstream channel length from the point at which the index is being calculated to the source of the stream (Fig. 4.7). In this study, a contour interval ΔH of 20 m is used for SL calculations. SL index values are in gradient meters. SL values of graded streams are constant, whereas relatively high values indicate steepening of the slope. Values of SL calculated for streams of the northern URG range from 1 to 500 (Fig. 4.8A).

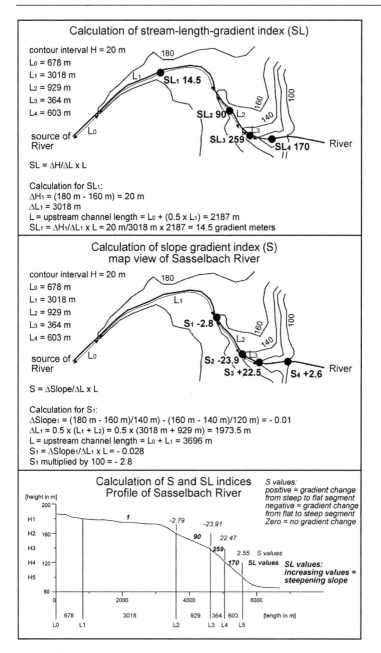

Figure 4.7. Sketch showing how the stream length-gradient (SL) and the slope gradient index (S) are calculated.

In this study, the first derivative of the SL index, the slope gradient index (S), is also used to describe changes in SL between segments and to show where the highest rates of SL changes are located. S is defined as

$$S = \frac{\Delta slope}{\Delta L} \times L \qquad\qquad Eq.\ 4.2$$

where $\Delta slope$ is the difference in slope between two adjacent segments next to each other. ΔL is the mid-point between the lengths of the two segments. L is the total channel length from the point of interest where the slope gradient index is being calculated to the source of the channel (Fig. 4.7). The dimensionless values of S can be positive or negative. For a constant gradient, S = 0, positive values indicate a gradient change from a steep to less steep segment, and negative values indicate an increase in segment steepness. The change from positive to negative S values and vice versa enables easier identification of subtle gradient changes than with the SL index. In this study, S reveals values between - 55.4 and + 120.3 (Fig. 4.8B).

4.4.1.2. Mountain-front sinuosity index

In order to assess the level of mountain-front faulting, the morphology of the mountain-front was investigated by calculation of the mountain-front sinuosity index S_{mf} (Bull and McFadden, 1977; Keller and Pinter, 1999; Burbank and Anderson, 2001). This index is frequently applied in regional studies on active tectonics (Wells et al., 1988; Marple and Talwani, 1993; Keller and Pinter, 1999; Burbank and Anderson, 2001; Cuong and Zuchiewicz, 2001; Silva et al., 2003) and reflects the balance between mountain-front faulting and the activity of transverse streams. The method is too insensitive to detect short term or minor fault reactivation that is insufficient to restore a straight front, or changes of the fluvial system due to a single climatic change. The mountain-front sinuosity is more of an indicator for the long-term level of faulting and erosional activity. In the URG, range-front faulting is extensional in style and tends to produce a straight front whereas the transverse fluvial system produces a wavy front shape by erosion and entrenchment into the front or by building up of fans along the front. The S_{mf} index is defined as

$$S_{mf} = \frac{L_{mf}}{L_S}$$ *Eq. 4.3*

L_{mf} is the length of the front along the foot of the mountain and L_s is the straight-line length of the mountain front. For the calculation, equal distances of L_s are used. L_{mf} is measured along the intersection of a horizontal plane that cuts the morphology at the foot of the mountain. Straight front segments give values of S_{mf} close to 1, whereas segments of low tectonic and relative strong erosional activity produce embayed fronts with $S_{mf} > 2$ or more (e.g. Burbank and Anderson, 2001). Calculations of the mountain-front sinuosity S_{mf} of this study have been carried out for the border faults and the scarp of the Oppenheim-Grünstadt Fault and yield S_{mf} values between 1.09 and 2.15 (Fig. 4.8C).

4.4.1.3. Valley shape index

The cross-sectional shape of valleys can reflect the rate of incision. This can be indicative of actively incising streams in areas of recent uplift, in particular in the footwall of mountain-front faults (border faults). The response to changes of the base level by faulting takes time to propagate upstream into the footwall. Measurements of the valley shape are therefore performed close to the border fault or mountain-front fault (e.g. 1 km upstream). Quantification of the valley-floor width-to-height ratio, V_f index, proved to be a useful tool to evaluate fluvial incision in uplifted areas (e.g. Bull and McFadden, 1977; Marple and Talwani, 1993; Cuong and Zuchiewicz, 2001; Azor et al., 2002; e.g. Silva et al., 2003). The V_f index is defined as

$$V_f = \frac{2 \times V_{fw}}{(E_{ld} - E_{sc}) + (E_{rd} - E_{sc})}$$ *Eq. 4.4*

V_{fw} is the width of the valley floor. $E_{ld} - E_{sc}$ is the elevation of left valley divide minus the elevation of the valley floor and $E_{rd} - E_{sc}$ is the elevation of the right valley divide minus the elevation of the valley floor. Low V_f values reflect V-shaped valleys of actively incising streams, which is commonly associated with uplift. High values indicate broad floored valleys of reduced incision in regions of relatively low uplift rates (e.g. Keller and Pinter, 1999). In addition to uplift, the shape of valley cross-sections depends on the lithology of the bedrock and the erosive ability of the river. V_f

measurements in the tectonically active Basin and Range Province of the U.S. Cordillera by Bull and McFadden (1977) provided values ranging between 0.05 – 47. V_f < 1 reflected steeply incised V-shaped valleys. V_f values of the URG range between 0.3 and 31.3 (Fig. 4.8D). The values < 2.2 indicate V-shaped valleys, though not as steeply incised as in the Basin and Range Province of the western USA, and values > 19.0 indicate significantly broad floored valleys.

4.5. Geomorphic index calculations for the northern Upper Rhine Graben

The results of the geomorphic indices calculations show that the values of the different indices vary significantly and point to regional differences that permit to separate the study area into regions based on value changes (Fig. 4.8). Table 4.1 summarizes mean and range of values for each geomorphic index in the individual regions.

Region	S index		SL index		Valley index		Sinuosity index	
	Mean of absolute value	Mean value	Mean value near border faults	Mean value	Range	Mean value	Range
Eastern graben shoulder								
Kraichgau	1.4	36	34 - 79	13.6	4.1 - 31.3	2.1	1.8 - 2.4
Odenwald	12.0	94	170 - 271	1.9	0.3 - 5.6	1.6	1.2 - 3.0
Sprendlinger Horst	0.4	31	16 - 33	19.9	14.7 - 27	1.9	1.5 - 2.5
Western graben shoulder								
WBF scarp	8.1	76	170 - 302	5.1	3.4 - 11.0	1.1	1.1 - 1.2
Rheinhess. Hügelland	7.8	74	71 - 101	2.4	1.1 - 4.6	1.4	1.1 - 1.8
Northern Pfälzer Wald	13.4	102	180 - 442	1.5	0.6 - 4.0	1.4	1.2 - 1.9
Southern Pfälzer Wald	1.3	52	49 -100	2.1	0.7 - 3.5	2.0	1.7 - 2.3
Northwestern regions								
Rhenish Massif	13.0	136					
Mainz Basin	2.6	63					

Table 4.1. Overview of the mean and range of values of each geomorphic index in the individual regions.

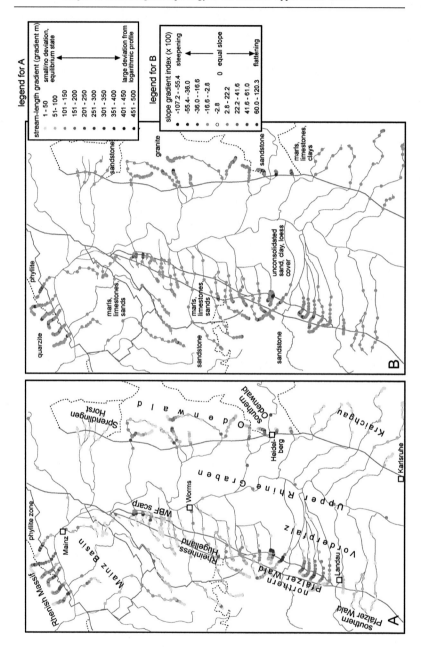

Figure 4.8. A) Stream-length gradients SL calculated for 1st and 2nd order streams that cross the border faults and the HTBF. Graded profiles are characterized by low SL (e.g. in the

Kraichgau). Concentrations of high values point to large deviations from graded profiles generally associated with steepening of the profiles. Stippled lines = lithological boundaries. B) Gradient changes S have been calculated for the same number of streams. Significant gradient changes are indicated by large value changes. Lithologies at surface are indicated.

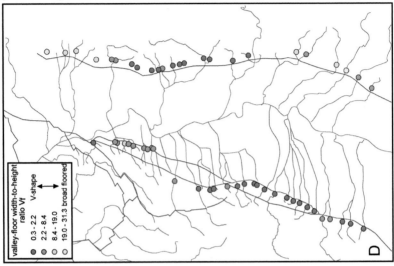

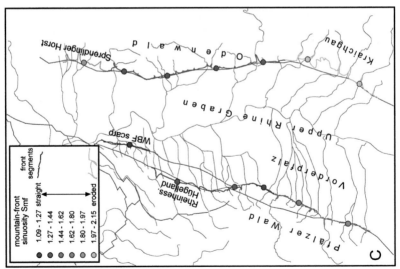

Figure 4.8. C) For the calculation of the mountain-front sinuosity S_{mf} front segments L_s of 15 km length are used. Values for S_{mf} are displayed in the center of the segments. D) Valley shapes have been calculated at 1 km distance from the border faults.

Eastern graben shoulder

For all indices calculated along the eastern graben shoulder, the Odenwald yields results that differ from those of the Kraichgau and Sprendlingen Horst (northern extents of the Odenwald; Fig. 4.8). The Kraichgau is characterized by lower relief as compared to the elevated areas of the Odenwald and the Black Forest (Fig. 4.1). The Odenwald and Black Forest have resistant basement rocks at surface whilst the Kraichgau has less resistant Mesozoic sediments at the surface (e.g. 1974a; Illies, 1974b). The streams draining the Kraichgau are graded streams with no or minor slope gradient changes (Fig. 4.8A, B). Note the very low values in Table 4.1. Further indications of erosional processes in the Kraichgau are the highly eroded fronts and the broad floored valleys. In combination with the low rock resistance and low topographic relief, the low and constant values of the gradient indices point to an equilibrium state of the streams and no indications for active uplift or incision in the Kraichgau. The Sprendlingen Horst is, similarly to the Kraichgau, characterized by a low relief in sandstones with eroded fronts and graded streams, which suggests higher erosional than tectonic activity.

Index values from the Odenwald differ from those of the Kraichgau and Sprendlingen Horst. North of Heidelberg, the Odenwald comprises basement rocks (mainly granites). To the south, sandstones are exposed. Index calculations of the mountain-front yield low sinuosity values, showing that the entire front from the sandstone to the granitic areas is almost straight. Valleys close to that front are dominantly V-shaped (Fig. 4.8D). Erosion of the mountain-front and valleys near the front are counteracted by the competent lithology (both granitic and siliciclastic). For this reason, the lithology of the bedrock rather than active tectonics is considered to control the shape of the mountain-front and valleys. In contrast to the index values along the mountain-front, gradient indices of river profiles draining the Odenwald are not uniform (Fig. 4.8A, B). Several smaller streams are not in equilibrium with uplift and incision. In particular, streams adjacent to the River Neckar, in the sandstone area, exhibit large gradient changes close to the mountain-front. Longer streams in the granitic Odenwald are in equilibrium with topography and do not show changes of slope gradients (Fig. 4.8B). Using the gradient measurements, the southern Odenwald is identified as a region dominated by tectonic processes that controlled local increases in shoulder uplift and graben subsidence.

Western graben shoulder

Index calculations show that the western graben shoulder can be divided into four regions (from N to S): the WBF scarp, the Rheinhessische Hügelland, and the northern and the southern Pfälzer Wald (Fig. 4.8). The Pfälzer Wald relief is developed almost exclusively in sandstone of Buntsandstein age. Given this overall similar composition, the large variations in geomorphic indices between the northern and southern part of the Pfälzer Wald is not related to rock resistance. All indices of the southern Pfälzer Wald suggest dominant erosional processes. Large gradient changes in the northern Pfälzer Wald correlate with a high topographic relief. Here, hilltops are approximately 100 m higher than in the south. The V-shaped valleys and the straight mountain-front of the northern Pfälzer Wald suggest active uplift that created relatively recently a higher topography. Furthermore, streams in this region show increased gradient changes along their courses in the Vorderpfalz, where they incise into Quaternary terraces (higher main and main terraces; Fig. 4.9, stream 3). To the north of the Pfälzer Wald, the topographic elevation difference between the shoulder and the URG, which is represented here by the southern Mainz Basin and the Rheinhessische Hügelland, reduces. Streams draining this area are relatively long, and uplift and incision are in equilibrium (Figs. 4.8B, 4.9, stream 4). The mountain front in this region has a medium level of erosion and valleys are wide. All indices indicate dominance of erosional activity. Longitudinal profiles of streams crossing the WBF scarp change from a nearly concave shape to a convex shape from south to north (Fig. 4.9, streams 5 - 8). Consequently, gradients (SL) and gradient changes (S), in particular close to the scarp front, increase northwards (Fig. 4.8B). The hanging valleys of the streams are the result of both bedrock changes and a non-equilibrium state, indicating that the streams have not yet adjusted to the topographic difference between the scarp and the floodplain of the URG. All valleys eastward to the WBF scarp are broad floored.

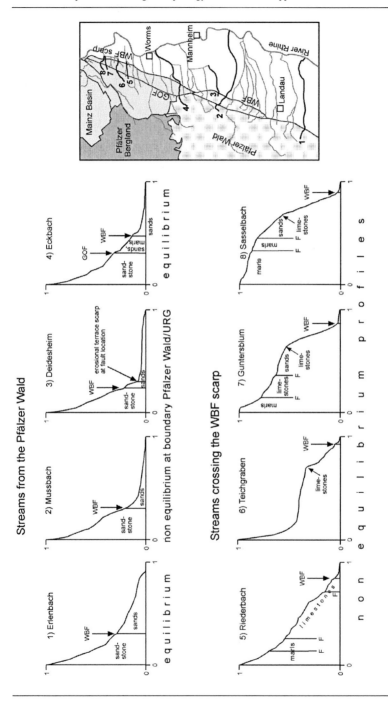

Figure 4.9. Normalized longitudinal profiles of streams crossing the Western Border Fault (WBF) and geological map. Near-surface lithology of the individual regions as in Figure 4.3: Pfälzer Wald – sandstone; Mainz Basin – marls, sands, limestones; URG – sands, loess cover. The position of the WBF and other faults (marked F) are indicated by the lines below the profiles. GOF - Grünstadt-Oppenheim Fault.

Southern Rhenish Massif and Mainz Basin

In the northwestern part of the URG, gradients of streams draining the Mainz Basin and the Rhenish Massif have been investigated (Fig. 4.8A, B). These two regions are structurally separated by the HTBF. Along strike of the fault lies the Mainz Bingen Graben. The HTBF is poorly expressed in the morphology and a distinct mountain-front is not developed (Fig. 4.1). For this reason, valley shapes and the mountain-front sinuosity have not been investigated. The concentration of high slope gradient changes in the southern Rhenish Massif contrasts low changes in the Mainz Basin. Streams draining the Mainz Basin erode into soft rocks (marls, sands, loess) of Late Tertiary to Quaternary age, whereas streams in the Rhenish Massif mainly erode into resistant Devonian rocks (quartzite, sandstones, siltstones). Consequently, the Mainz Basin exhibits gentle slopes and the Rhenish Massif steep slopes. These observations strongly suggest that the gradient changes reflect the differences in rock resistance and relief. At a large scale, the topographic difference between the Rhenish Massif and the Mainz Basin are related to the greater tectonic uplift of the Rhenish Massif during Quaternary times. The local anomalies of gradient changes of streams in the southern Rhenish Massif occur at various locations along the streams and may be related to local lithology changes. There are no significant indications of recent activity of the HTBF.

4.6.   Discussion

4.6.1.   Interpretation of geomorphic indices

The results of this geomorphological study show that areas along the border faults of the northern URG are characterized by a tectonic morphology with narrow and partly hanging valleys and straight mountain-fronts. This suggests a segmentation of the border faults, with higher levels of tectonic activity located along the WBF scarp, the northern Pfälzer Wald and the southern part of the Odenwald, and lower levels (associated with landforms shaped by denudational processes) in the southern Pfälzer Wald, the Kraichgau and the northern Odenwald (Sprendlinger Horst; Fig. 4.8).

The geomorphic indices calculated for the southern Odenwald and northern Pfälzer Wald show that these areas have a topography strongly influenced by tectonic activity. It is interesting to note that the Odenwald and northern Pfälzer Wald are characterized by the highest differences in relief between the shoulder and the graben, resulting in the straight mountain-fronts. The *Heidelberger Loch*, which was during Tertiary graben evolution and particularly during the Quaternary a locus of deposition (~ 380 m of Quaternary deposits), is associated with the Eastern Border Fault. It is situated adjacent to the highest topography of the Odenwald, very close to the mountain-front (566 m elevation of the Odenwald east of Heidelberg). In this context, it is reasonable to suggest that this part of the Eastern Border Fault along the southern Odenwald has been active recently, and contributed to the significant uplift of the shoulder and the subsidence of the graben. Unfortunately, similar independent geological and geomorphological data is not as conclusive for the northern Pfälzer Wald , which has fewer indicators of active tectonics. The extensional faulting of Riss loess at the outcrop at Forst (outcrop marked 3 in Figure 4.5-5) is one significant indicator. In addition, terrace widths adjacent to the WBF (higher main and main terraces) are much smaller than the same terrace levels further south (Figs. 4.5-3, 4.5-4). This may be a second indicator, since smaller terraces would form during periods of increased uplift of the northern Pfälzer Wald and contemporaneous graben subsidence.

The WBF scarp with hanging, dry valleys and a linear scarp front attest to a morphology that developed in response to both tectonic and erosional processes (Figs. 4.9, 4.10). The linear front can be explained by lateral scarp erosion and terrace formation, with the flight of middle terraces constituting the scarp (*Chapter 3*; Fig. 4.10). Late Pleistocene to Holocene faulting activity on the WBF has been documented from a trench site at the southern end of the WBF scarp (*Chapter 2*). This suggests that lateral scarp erosion was accompanied by WBF activity and that both erosional and faulting processes occurring during the Pleistocene resulted in the formation of the 20 km long scarp. The scarp height (50 – 100 m) is the result of fluvial incision counteracting regional uplift of the Mainz Basin. Neither minor faulting activity (0.5 m; *Chapter 2*) nor erosion has modified the scarp morphology significantly since fluvial activity at the scarp ceased at the end of the Late Pleistocene. A decrease in fluvial erosion in the Holocene is indicated by the small size of the alluvial fans covering the lower terrace at the foot of the scarp. These fans are too small to constitute the eroded material of the wide valleys (Fig. 4.10). Thus, most of the erosion of the valleys must have occurred prior to the Holocene.

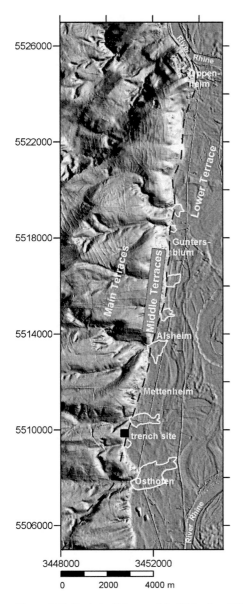

Figure 4.10. Shaded relief map of the 20 km long WBF scarp using DEM data from DGM5 (2002). The surface trace of the WBF (dashed line) follows the base of the scarp (Steuer, 1911;

Franke, 2001). The outlines of alluvial fans are mapped after Franke (2001) and topography data. The mapping of terraces is after Chapter 3.

Recognizing a tectonic morphology in the present-day landscape does not necessarily imply that tectonic activity is currently still ongoing. Tectonic morphology may have developed long ago and has been preserved since. Likewise, denudational and dissected morphology can be preserved long after erosion and sedimentation processes cease. The geomorphological evolution of the northern URG (section 4.2.4) shows that the morphology of the graben shoulders evolved over the entire Pleistocene. Fluvial processes (erosion, sedimentation, incision) were most effective during glacial periods when erosion formed wide valleys in many parts of the shoulder areas. Since the end of the Pleniglacial approx. 15,000 years BP, erosional activity has decreased significantly and the landscape has changed little (e.g. Brüning, 1975; Klaer, 1977). Wide Pleistocene valleys are presently preserved as dry valleys. While erosion decreased, the low level of faulting activity remained unchanged. Deformation rates inferred from displaced Early to Middle Pleistocene terraces that were deposited along the western border of the URG, in the Mainz Basin and in the Lower Main Valley, suggest a long-term deformation rate of 0.01 – 0.08 mm/yr (*Chapter 3*). Displaced deposits of Late Pleistocene to Early Holocene age yield a rate in the order of 0.03 mm/yr, close to the mean value for the Early to Middle Pleistocene. These young displacements have been observed on intra-graben faults (Semmel, 1969; Monninger, 1985; *Chapter 2*) and on the WBF itself (*Chapter 2*). Since erosional and depositional activity has been lower than tectonic activity during the last 15,000 years, the tectonic morphology of the northern URG has been preserved until present. It is argued that this tectonic morphology reflects long-term tectonic processes.

4.6.2. Interpretation of structural control on the landscape

The structural setting of the northern URG has exerted a significant control on the morphology of the graben borders, which are currently expressed by straight, steep mountain-fronts. Intra-graben faults have affected the positions of the River Rhine and its tributaries, as documented by several fluvial terrace scarps situated along faults (composite scarps marked C on Figure 4.6A). Faulting activity is considered responsible for the alignment of the terrace scarps and increased incision in the hanging wall, leading to increases in the heights of these composite scarps. It is suggested that the youngest composite scarp formation occurred in the southern part of the study area along the eastern and western border of the recent floodplain. At scarps west of Karlsruhe and at the western scarp south of Mannheim the lateral extents of the Holocene Rhine meanders are constrained by faults, giving the scarps a linear character for distances of approx. 15 km.

Comparison of fault and stream orientations suggests recent structural control on small streams in the southeastern part of the graben and on the course of the Late glacial River Neckar. Several rivers of the Mainz Basin follow linear valleys that seemed to have been caused by fault activity (Wagner, 1930; this study). Since there are no indications for a recent structural control on rivers of the Mainz Basin, its Pliocene and Early Pleistocene tectonic activity was probably more significant, the result of which is still observed today (Figs. 4.5, 4.6, 4.8A, B).

The calculations of slope gradient changes show that, particularly at the crossing of the border faults, significant gradient changes in stream profiles occur where rivers cross faults (Figs. 4.8A, B). Consideration of all the geomorphic indices calculated allows identification of the active segments of the border faults (Fig. 4.11). Measurements along stream profiles show that graded stream profiles dominate the URG valley. Measurement results suggest that intra-graben faults exert a minimal control on the rivers. None of the intra-graben faults, which have been detected as active in comparison with morphology and drainage orientations, caused gradient changes on the rivers (Figs. 4.6, 4.8A, B). The only exception is the composite scarp close to the WBF along the northern Pfälzer Wald. These contradictory observations may be related to the resolution used for gradient measurements. The effects of intra-graben faults may be sufficiently subtle that only the use of a higher resolution DEM and smaller contour intervals (< 20 m) may detect small gradient changes of river courses within the valley.

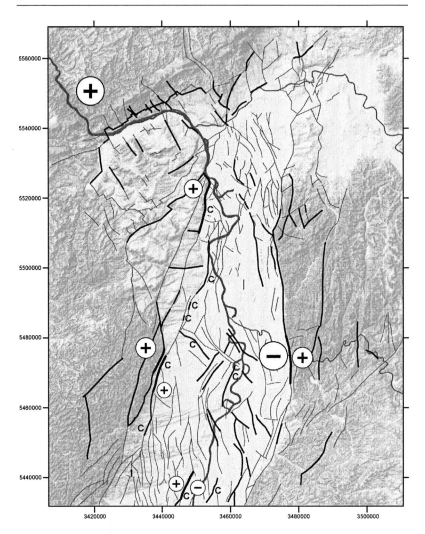

Figure 4.11. Summary of the key interpretations of this study. The relative vertical movements refer to the most recent motions of Late Pleistocene age inferred from the paleogeographic evolution and geomorphic indices calculations. Grey lines show the entire fault network. Faults marked in black provided indications for long-term structural control on the landscape formation and the drainage system during the Quaternary. Fault segments that have been active recently (~ Late Pleistocene to Early Holocene), are indicated with thick solid lines. C = scarp attributed to combined tectonic and fluvial activity, + uplifting areas, – subsiding areas.

4.7.   Conclusions

By synthesizing the available geological and geomorphological data in a series of paleogeomorphic maps (Fig. 4.5), constructing overlays of topographic, fluvial and structural features (Fig. 4.6) and calculating geomorphic indices (Fig. 4.8) this study provides a regional overview on the activity on the URG Border Faults (Fig. 4.11). Fault segments with indicated increased tectonic activity are located along the WBF scarp, the northern Pfälzer Wald and the southern part of the Odenwald. This activity has resulted in vertical displacements, of which the largest occurred on the Eastern Border Fault adjacent to the *Heidelberger Loch*. About 380 m of Quaternary down faulting is documented for this fault segment. The highest topography of the footwall close to the border fault is also located adjacent to this segment, suggesting a significant amount of Quaternary shoulder uplift (Fig. 4.11). In addition to the border faults, numerous faults in the study area exhibit apparent morphological expressions in the form of scarps and valleys. This observation leads to the conclusion that activity on these faults had a significant impact on the landscape. Pleistocene tectonics has been counteracted by the dynamics of the River Rhine system. The resulting fluvial terraces bear fault alignments and, at active boundary faults, the terrace flights form composite scarps of considerable height (Fig. 4.11).

The rate of tectonic deformation in the northern URG has been at a low level for the entire Pleistocene (~ 0.03 mm/yr), while denudational processes increased during glacial periods. During the last 15,000 years, erosional and depositional activity has decreased drastically whilst tectonic activity remained constant. The tectonic morphology now preserved in the northern URG was created by long-term, low level tectonic activity and has been preserved because of the decrease in erosive activity in the last 15,000 years.

CHAPTER 5

FAULT REACTIVATION ANALYSIS (SLIP TENDENCY) OF UPPER RHINE GRABEN FAULTS

5.1. Introduction

At present, the Upper Rhine Graben (URG) area is among the most seismically active areas north of the Alps, with activity levels generally low to moderate (Fig. 1.8). A wide distribution of earthquake epicenters occurs in the southern part of the URG and the adjacent shoulder areas. In the central and northern URG and its shoulder areas only few scattered events have been reported (Ahorner, 1983; Bonjer, 1997; Leydecker, 2005a; Fig. 5.1). Focal depths (relative to the surface) in the URG range from 5 to 20 km with the majority of events taking place in the upper 10 km in the basement rocks underlying the graben fill (Baier and Wernig, 1983; Bonjer et al., 1984). The largest, damaging earthquakes occurred at the southern border of the graben in the Basel area and to the east of the Black Forest in the Swabian Alb: Basel, 1356; M_w 6.0 – 6.6 (Ahorner and Rosenhauer, 1978; Mayer-Rosa and Cadiot, 1979; Mayer-Rosa and Baer, 1992) and Albstadt, 1978; M_L 5.6; (Leydecker, 2005a). In the central graben segment, three damaging earthquakes with local magnitudes not exceeding 5.4 have been documented (Leydecker, 2005a). In the northern URG, only two damaging earthquakes are recorded (Fig. 5.1). Documented earthquakes in the URG have not been large and did not produce a surface rupture. For the lack of large earthquakes, several explanations have been proposed. The seismic catalogue is either incomplete and does not cover the long recurrence intervals of typical intra-plate earthquakes. Alternatively, frictional resistance on the generally steeply dipping faults might be high enough to withstand shear stresses (Illies et al., 1981). A more conservative approach suggests that due to the high fracturing of the URG area sufficient energy for large earthquakes cannot accumulate but releases through smaller events (M < 5) or creeping movements (Ahorner, 1975; Bonjer et al., 1984; Fracassi et al., 2005).

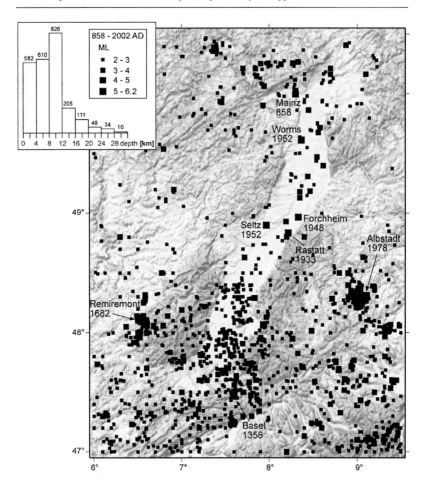

Figure 5.1. Seismicity of the URG area. Catalogue after Leydecker (2005a). For conversion of I_o to M_L the formula $M_L = 0.636*I_o + 0.4$ of Rudloff and Leydecker (2002) has been used. The histogram displays the distribution of focal depths for the earthquakes recorded between 858 and 2002 AD (N=2943). The majority of earthquakes have documented depths of 8 to 12 km. Note that the large number of shallow focal depths (< 4 km) includes also historical earthquakes that commonly have been assigned with a depth of 0 km.

The high numbers of pre-existing faults in the URG area represent zones of weakness. Despite the lack of large earthquakes presently observed, the occurrence of damaging earthquakes in the future cannot be fully excluded and thus these faults have to be considered potential sources for future earthquakes. Therefore, assuming that the present-day stress field has a similar orientation to those identified throughout the grabens' evolution, it is more likely to reactivate an existing fault than to create a new one. A detailed understanding of fault reactivation is thus of great economical and societal importance. In addition, oil and gas exploration considers faulting activity in order to evaluate bore hole stabilities and the risk of leakage. In order to assess the reactivation potential of faults, information on the state of stress, fault geometries, pore fluid pressure and material properties (both of the fault and the unfaulted rock) are required. For the URG, fault geometries are best constrained from geological and geophysical investigations. Stratigraphic well correlations and seismic profiling provide information on the magnitude of vertical throws on the border faults and intra-graben faults and enable to identify faults that have been frequently reactivated during the graben evolution (e.g. Tietze et al., 1979; Derer et al., 2003; Haimberger et al., 2005). Geomechanical finite element (FE) modeling is a tool that can provide detailed constraints on estimates of in-situ stress magnitudes and orientations of a geological system as well as the behavior of faults and fault properties. For this purpose, recent 3D FE modeling of the URG was performed (Schwarz and Henk, 2005; Buchmann and Connolly, 2007). The FE models of Buchmann and Connolly (2007) and Schwarz and Henk (2005) cover the entire URG and include the border faults and major fault structures (e.g. Hunsrück-Taunus Boundary Fault, HTBF). The reactivation of pre-existing faults is strongly dependent on the material properties (mainly friction) of the faults. One important result of the modeling of Schwarz and Henk (2005) is that the observed displacement pattern of the URG could only be reproduced when using low friction on the border faults (coefficient of friction μ of 0.3 – 0.4). Buchmann and Connolly (2007) present a stress field model that is calibrated with all available in-situ stress data from the URG. Their results show that with higher friction values ($\mu = 0.6$) realistic fault displacements can be achieved.

In this study, the FE approach is used to simulate the three-dimensional state of stress with the aim of assessing the reactivation potential of major faults in the URG under the present-day state of stress. The study comprises two modeling parts: (1) 3D stress field modeling and (2) 3D slip tendency (ST) analysis of faults in the study area. The 3D stress field obtained from the FE model is used to calculate the ST parameter on major faults. The ST parameter, introduced by Morris et al. (1996), is a quantitative measure of the likelihood of fault reactivation and is used to identify potentially sliding fault segments in a given stress field. Since the 3D stress field is used as the basis for the ST analysis, emphasis is placed on simulating the stress field as realistically as possible. The investigation of ST is performed for both intra-graben faults and faults of the shoulder areas. Thus, in contrast to the other studies of this thesis, which focused mainly on the northern URG (*Chapters 2 to 4*), the area investigated for ST includes the entire URG and adjacent graben shoulders. This fault analysis is extended to a larger area in order to compare the fault behavior of the northern URG to the central and southern parts of the graben and surrounding regions. The fault reactivation predicted by modeling is compared to existing data on active faulting. Emphasis is placed on comparing the ST results with active faults of the northern URG that have been identified in the framework of this thesis by paleoseismological trenching and geomorphological methods. The modeling of fault reactivation aims also to investigate a possible correlation between areas of high slip potential and the distribution of seismicity. In addition, fault segment orientations with high ST and orientations of nodal planes of earthquake focal mechanisms are compared.

5.1.1. Contemporary stress field and kinematics of the Upper Rhine Graben

The present-day URG stress field has been the subject of several seismological studies inverting a stress tensor from earthquake fault plane solutions (Ahorner et al., 1983; Larroque et al., 1987; Delouis et al., 1993; Plenefisch and Bonjer, 1997; Hinzen, 2003). The studies suggest regional variations in the stress field. For the southern URG, inversions of the principal stresses reveal that the region is transtensional, with the strike-slip component being slightly more dominant (Plenefisch and Bonjer, 1997). The northern URG is characterized mainly by an extensional tectonic regime with a less dominant strike-slip regime (Delouis et al., 1993; Plenefisch and Bonjer, 1997; Hinzen, 2003). Kinematic analysis of fault slip data from outcrops in the graben shoulders adjacent to the border faults revealed predominantly strike-slip faulting and minor oblique-slip to pure extensional faulting (Lopes Cardozo and Behrmann, 2006). This is interpreted as indicating that the border faults are characterized by sinistral transtension.

A large number of in-situ stress determinations are available from the scientific geothermal drilling site at Soultz-sous-Forêt located in the central part of the URG (Fig. 5.2). The wells at this site reach depths relative to the surface of 3 to 5 km (Cornet et al., 2007). For these depths, data of stress orientations, stress magnitudes and rheological parameters is available, which is a very valuable dataset to be used for calibration of the FE model of this study. The stress regimes at the drilling site were determined mainly based on hydraulic fracturing stress measurements and are of an extensional type for the upper 3 km of the crust and strike-slip below (max. measurement depth 3.5 km; Rummel and Baumgärtner, 1991; Klee and Rummel, 1999). These stress conditions are in agreement with the results from earthquake focal mechanisms for the URG, which suggest a stress regime in transition from extension to strike-slip. However, it should be noted that from the earthquake data the stress regimes could not be differentiated with depth (Bonjer, pers. comm. 2004). In summary, all stress field investigations point to a transitional stress field between strike-slip and extension both for the URG and the adjacent shoulder areas. At a large scale, the NNE-SSW trending graben structure is subjected to sinistral transtension under the present-day stress field. This stress field is stable since the beginning of the Miocene (Illies, 1975; Schumacher, 2002; Michon et al., 2003; Dèzes et al., 2004; Fig. 1.2).

The direction of the maximum horizontal stress component S_H in the URG area is on average NW-SE (N150°) as demonstrated in the World Stress Map (WSM, Reinecker et al., 2004; Fig. 5.2). The WSM is a database of stress orientations and stress magnitudes that have been determined from focal mechanisms, borehole breakouts, drilling-induced fractures, in-situ stress measurements such as hydrofractures, overcoring and borehole slotter, as well as geologic data from fault slip analysis and volcanic vent alignments. In-situ stress measurements at the Soultz-sous-Forêt site revealed on average a NNW-SSE (N170±10°) orientation of S_H. In several wells, a rotation of the S_H orientation from NW-SE to NNW-SSE was observed with increasing depth (Cornet et al., 2007). This rotation towards the north with depth was also documented for the KTB drilling site in SE Germany, located on the western flank of the Bohemian Massif some 300 km to the ENE of Soultz-sous-Forêt (Brudy and Zoback, 1999), and from a geothermal well in Bad Urach, some 130 km to the SE (Heinemann et al., 1992; Tenzer et al., 1992; Fig. 5.2). In all three wells, the average orientation of S_H at a depth of 3 km was N170±15°. It remains uncertain why this is about 20° north to the mean S_H orientation for Central Europe (N150° from WSM database; Cornet et al., 2007)).

Several authors state that the stress regimes and the present-day velocity field of the URG area reflect the plate boundary loads (e.g.Müller et al., 1992; Plenefisch and Bonjer, 1997). The effects of ongoing African-Eurasian convergence on areas north of the Alps are currently not well constrained because GPS measurements have not occurred sufficiently long to accurately determine the small intraplate motions (Vigny et al., 2002). Based on GPS measurements and modeling studies on the GPS velocities currently available, the crustal motions of the URG area are characterized by a NW directed push in the order of 1 mm/year and an E-W opening of the graben of 0.5 mm/year (Vigny et al., 2002; Nocquet and Calais, 2004; Rózsa et al., 2005; Tesauro et al., 2005). These results are consistent with the strain determined from seismological studies and fault slip analysis that show a NW-SE shortening (S_H) and NE-SW extension (S_h minimum horizontal stress component) in the URG area (Plenefisch and Bonjer, 1997; Lopes Cardozo and Behrmann, 2006).

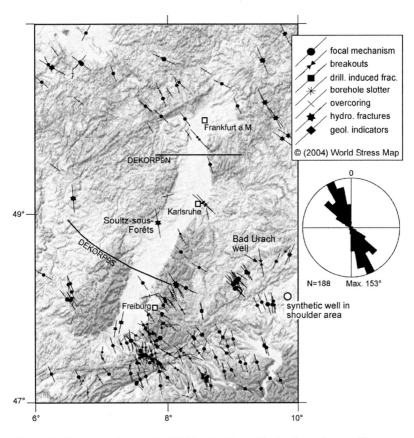

Figure 5.2. Stress map showing the NW-SE orientation of S_H for the study area. The stress orientations are relatively homogenous and only few stress perturbations occur, most of which lie in the southern URG. The rose plot has been created using World Stress Map data of all

listed methods with qualities A to C of the area displayed. Database: World Stress Map (Reinecker et al., 2004). DEKORP9N and DEKORP9S are two seismic reflection lines shown in Figure 5.3.

5.1.2. Crustal structure of the Upper Rhine Graben area

Information on the crustal structure of the URG area comes from two seismic reflection lines shot across the URG by the French and German Deep Reflection Seismic Consortiae in 1988 (ECORS and DEKORP; Brun et al., 1991; Wenzel et al., 1991). The DEKORP9N line crosses the northern part of the URG and the DEKORP9S line crosses the southern URG (Fig. 5.2). Both seismic lines resolve the entire crust and parts of the upper mantle (Brun et al., 1991; Wenzel et al., 1991; Fig. 5.3). The profiles reveal a thickness of 16 to 18 km for the upper crust. The crust-mantle boundary (Moho) lies at approximately 27 km depth in the northern URG and 24 km in the southern URG. The 24 km thin crust is the thinnest continental crust north of the Alps (Dèzes and Ziegler, 2002; Dèzes et al., 2004). The dome-shaped structure of the Moho topography in the southern URG is also referred to as Vosges-Black Forest Dome (or Vosges-Black Forest Arch; e.g. Dèzes et al., 2004).

The upper crust of the URG area is highly fractured and has been imaged by numerous seismic surveys undertaken in the 1960s and 1970s. The two deep reflection seismic lines show the traces of the graben border faults relatively well for the shallow parts of the upper crust; their extension to greater depth is less well constrained (Fig. 5.3). It is assumed that the border faults reach to a depth of at least ~15 km. Faults in the basement underneath the graben extend to a maximum depth of ~10 km (Brun et al., 1991; Wenzel et al., 1991; Fig. 5.3). The depth of 15 km coincides with the lower depth of seismicity under the graben (Leydecker, 2005a) and the limit of brittle deformation (Bonjer et al., 1984; Plenefisch and Bonjer, 1997). Exceptional is the southeastern part of the URG, where hypocenters at or beneath the eastern shoulder of the URG reach almost to the crust-mantle boundary at a depth of 25 km (Dinkelberg area east of Basel; Bonjer, 1997). In this region, the lower crust is thinned and characterized by episodic brittle failure (Mayer et al., 1997), while for the rest of the URG area ductile deformation can be assumed to occur in the lower crust (Brun, 1999).

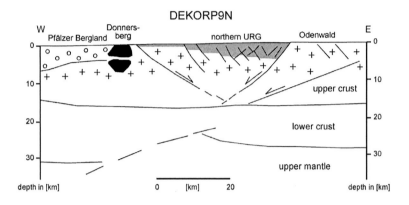

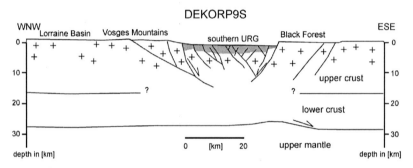

Figure 5.3. Geological interpretations of the seismic reflection profiles DEKORP9N (after Wenzel et al., 1991) and DEKORP9S (after Brun et al., 1991) shot across the URG. The existence of low angle shear zones in both profiles (indicated with arrows) as well as the thickening of the lower crust in the western part of profile DEKORP9N are a matter of ongoing debate (Zeis et al., 1990; Henk, 1993; Schwarz and Henk, 2005). For location of profiles see Figure 5.2. Grey = graben fill, black = rhyolite, circles = Permo-Carboniferous rocks.

5.1.3. Active faults in the Upper Rhine Graben area

Figure 5.4 shows the main faults of the URG and the shoulder areas. The seismicity of the region is widespread and only few fault strands could be identified as seismically active (Figs. 1.5, inset I and 5.4):

- a graben-parallel fault south of Rastatt in the central graben segment (active during Rastatt 1933 earthquake; Ahorner and Schneider, 1974; Fracassi et al., 2005),

- the Sierentz Fault west of Basel (Sierentz 1980 earthquake; Lopes Cardozo et al., 2005),

- the Eastern Border Fault north of Freiburg (Mahlberg/Lahr 1728 earthquake), and
- small fault strands in the southern and central URG (Kaiserstuhl 1926, Offenburg 1935 and Wissenbourg 1952 earthquakes; Fracassi et al., 2005).

Geological observations provide also information on fault segments that have been active during the Quaternary. The map in Figure 5.4 shows a compilation of active faults. Mapping of these faults is based on 2D seismic profiles recorded on the River Rhine, which enabled identification of vertical displacements in Quaternary sediments on intra-graben faults (displacements in the order of tens of meters; (Haimberger et al., 2005; G. Wirsing, pers. comm., 2005). Active segments of the border faults are mapped after Illies (1975). It should be noted, however, that this author did not give details on his interpretation. The map in Figure 5.4 is supplemented with results of this study from the northern URG, which provided information on active fault segments based on paleoseismological trenching, geomorphological interpretations and terrace mapping (*Chapters 2, 3 and 4*). The compilation highlights that little information is known on active faults in the shoulder areas. One main target of this ST analysis is to investigate the reactivation potential of the active fault segments of Figure 5.4.

Next page:

Figure 5.4. Fault segments in the URG area with documented Pleistocene activity. The faults are compiled after: Breyer and Dohr (1967), Ahorner and Schneider (1974), Illies (1975), Cushing et al. (2000), Brüstle (2002), Fracassi et al. (2005), Haimberger et al. (2005) and G. Wirsing pers. comm. (2005). The large number of Pleistocene faults in the northern URG result from the compilation of this study using paleoseismological and geomorphological methods (see Chapters 2, 3 and 4 and Figures 2.4, 3.2A, 4.11 herein). Faults with no age constraint after Andres and Schad (1959), Straub (1962), Behnke et al. (1967), Breyer and Dohr (1967), Illies (1967; 1974a), Tietze et al. (1979), Stapf (1988) and Derer (2003). RA = Rastatt.

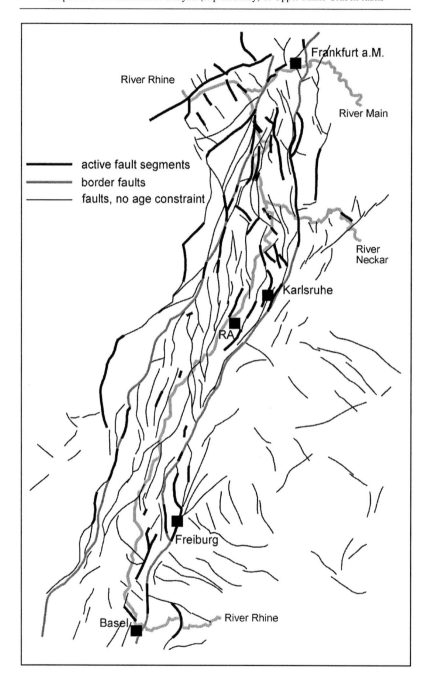

5.2.   Finite element modeling of fault reactivation

5.2.1.   Introduction to finite element modeling

Finite Element (FE) modeling is a method suitable to investigate the state of stress of a three-dimensional geological system, fault displacements and the reactivation potential of fault structures (e.g. Dirkzwager, 2002; Connolly et al., 2003; Zhao and Müller, 2003; Cornu and Bertrand, 2005; Schwarz and Henk, 2005; Ellis et al., 2006; Buchmann and Connolly, 2007). The FE technique was developed for mechanical engineering to solve complex problems, where no analytical solution was possible.

FE modeling is a method that enables to solve continuum problems. The continuum is an aggregation of idealized particles (finite elements) with specific properties. A wide range of finite elements has been developed for most physical and chemical situations. In 3D geomechanical modeling hexahedral, tetrahedral and other prismatic shaped elements are most useful. The geometry of the elements is defined by the corner points of the elements i.e. nodes. More complex second order elements contain additional nodes on the sides. In FE modeling, physical properties and constraints, defining degrees of freedom for displacements, are assigned to the elements, which then act on the nodes. Additional loads may be placed on element faces or within the elements. A model solution is produced by achieving equilibrium of forces for each element and by following principal laws of continuum mechanics. This includes construction of a system of simultaneous equations that is solved for unknown values using matrix algebraic methods with either linear, nonlinear or x^{th} order partial differential equations. The most common approach in FE modeling of a geological system is the 'displacement method' in which the displacement is the unknown term. This method is based on the concept that element displacement is defined as a function of nodal displacement, which means that when the nodes displace, they drag the elements along. From the displacement, strain and subsequently stress can be determined and hence the stress state of the system modeled can be defined. For a complete description of the FE method the reader is referred to Zienkiewicz et al. (2005). A summary of FE analysis aimed at Earth scientists is available in Ramsay and Lisle (2000).

For the numerical modeling presented in this study, mainly commercial software was used. This includes the finite element code ABAQUS^{TM}, the commercial pre-processing software HYPERMESH^{TM} and the post-processing software GeoMoVie[1] for 3D stress, strain and displacement analysis. The FE modeling results obtained from ABAQUS include the stress- and strain tensors at the element's integration points and the displacements at the nodes. In order to visualize various derivatives of the stress field (e.g. maximum horizontal stresses, differential stress) as well as tectonic regimes, the output data from ABAQUS needs to be post-processed, which is performed with GeoMoVie. With this software, the slip tendency parameter is also calculated on the fault planes. For visualization of the various results, GOLDEN SOFTWARE Surfer8 is used.

5.2.2. Theory of fault reactivation

Slip of a pre-existing fault surface can be described with a simple constitutive relationship, referred to as Amontons's Laws from 1699 (based on observations by Leonardo da Vinci; revised theory in the modern publication of Jaeger, 1969). This relationship states that the shear stress necessary to initiate slip on a surface is proportional to the normal stress across the surface. The proportionality constant is the coefficient of friction μ:

$$\tau = \mu\sigma_n \qquad\qquad Eq.\ 5.1$$

τ – magnitude of shear stress

σ_n – magnitude of normal stress

The coefficient of friction μ is then defined as the ratio of shear stress to normal stress:

$$\mu = \frac{\tau}{\sigma_n} \qquad\qquad Eq.\ 5.2$$

Three types of friction can be discriminated. μ_i is the internal friction required to create a rupture surface, the static friction μ_{static} is required to initiate sliding on an

---

[1] GeoMoVie is freely available at http://www-wsm.physik.uni-karlsruhe.de/pub/news/news_frame.html

existing surface and the dynamic friction $\mu_{dynamic}$ represents the apparent friction on a surface during the process of sliding.

Currently, no friction law is available that describes all micromechanical processes and the complex interactions of surface contacts that evolve during the sliding process. Therefore, friction laws are phenomenological descriptions of the friction behavior and are based on empirical studies (for an overview of friction laws and friction experiments see Scholz, 2002). Values for the coefficient of static friction μ_{static} for cohesionless faults are provided by empirical data obtained from friction experiments on a variety of rock types. The principal finding of these experiments is that, with the exception of several clay minerals, friction is independent of rock type. For the upper crust the friction coefficient is independent of cohesion and rather homogenous (Byerlee, 1978). Depending on confining pressure, Byerlee (1978) describes two relationships, commonly quoted as "Byerlee's law":

$$\tau = 0.85\sigma_n \qquad Eq.\ 5.3$$

for pressure conditions with $\sigma_n < 200$ MPa, and

$$\tau = 50 + 0.6\sigma_n \qquad Eq.\ 5.4$$

(in MPa) for $\sigma_n > 200$ MPa. Values for μ_{static} can be lower than 0.6 if fault surfaces contain fault gouges with clay minerals (Byerlee, 1978).

Values for the coefficient of static friction may also be calculated by the inversion of earthquake focal mechanism data. Using this method, friction values for μ_{static} are much more diverse and commonly less than the 0.6 one would use if applying Byerlee's law blindly. For example friction values determined from focal mechanisms of the southern URG are in the range of 0.3 to 0.6 with few exceptions indicating friction even as low as 0.1 (Plenefisch and Bonjer, 1997). For a number of reasons the friction law of Byerlee maybe applicable as a base case when estimating the strength of natural faults. The law is independent of lithology, sliding velocity and the roughness of fault surfaces; it is valid over a wide range of hardness and ductility of various rock types and temperatures up to 350°C for silicates (summarized in Scholz, 2002, p. 67, see references herein). In addition, the analysis of earthquakes with unambiguous fault planes (Jackson, 1987; Sibson and Xie, 1998) and in situ stress measurements of deep (> 1 km) boreholes in a variety of tectonic regimes (McGarr and Gay, 1978; Zoback and Healy, 1984; Zoback and Healy, 1992; Brudy et al., 1997) confirm that active faults exhibit coefficient of friction values in the range of 0.6. These are similar to those observed in laboratory experiments. From the Soultz-sous-Forêt borehole data, slightly higher values in the range of 0.8 for μ_{static} have been obtained (Cornet et al., 2007). The KTB borehole in southeast Germany delivered a value of 0.6 (Brudy et al., 1997).

5.2.3. Theory of slip tendency

Amontons's Laws (Equation 5.1) describes the critical stress condition for sliding as a linear relationship between shear and normal stresses. Following this theory, the shear stress to normal stress ratio has been introduced by Morris et al. (1996) as the slip tendency parameter T_S:

$$T_S = \frac{\tau}{\sigma_n} \qquad\qquad Eq.\ 5.5$$

This parameter is used as a quantitative measure of the likelihood of fault reactivation and to identify potentially sliding fault segments in a given stress field. The maximum value of T_S is limited by the slope of the sliding envelope in the Mohr diagram. This means that sliding occurs in the case T_S is greater than the coefficient of static friction μ_{static} on the fault surface, assuming that this fault is cohesionless:

$$\text{Sliding} = T_S \geq \mu_{static} \qquad\qquad Eq.\ 5.6$$

In this study, a normalized index of slip tendency ST is used with a range always between 0 and 1. This is found by dividing T_S by its maximum possible value, which is the value of μ_{static}:

$$ST = \frac{\tau}{\mu_{static}\sigma_n} \qquad\qquad Eq.\ 5.7$$

Slip occurs when the slip tendency becomes equal 1 or larger. However, Morris et al. (1996) showed several recent examples and argued that slip tendency values of 0.7 and larger indicate that slip is likely to occur. This highlights that for the interpretation of slip tendency results relative changes of slip tendency along faults are relevant and not so much absolute values. A higher slip tendency implies that the state of stress along a given fault surface is closer to failure and that consequently the likelihood of slip is higher.

In a more complex scenario, failure depends furthermore on the cohesive strength of the fault surfaces and the pore fluid pressure:

$$\tau = C + \mu_{static}(\sigma_n - P_f)$$ *Eq. 5.8*

$$\mu_{static} = \frac{(\tau - C)}{(\sigma_n - P_f)}$$ *Eq. 5.9*

$$ST = \frac{(\tau - C)}{\mu_{static}(\sigma_n - P_f)}$$ *Eq. 5.10*

P_f – magnitude of pore fluid pressure

C – magnitude of cohesion

Several studies (e.g. Reches, 1987; Krantz, 1991; Twiss and Moores, 1992; Zoback and Healy, 1992) have demonstrated that cohesion along fault surfaces decreases as subsequent slip occurs. Faults with large offsets and a long history of activity generally have a low cohesion. Therefore, neglecting cohesion along pre-existing fault surfaces is a reasonable simplification. Furthermore, neglecting cohesion in a slip tendency analysis yields a conservative approach since it overestimates the slip tendency of faults that are cemented or have an otherwise apparent cohesive strength (Streit and Hillis, 2004). At crustal scales, the faults in the URG area have a long history and there are no indications for fluid pressures in excess of hydrostatic pressure. Hydrostatic pressure is a simple function of depth and is therefore incorporated in σ_n. Given these two arguments the approach used in this study is based on Equation 5.7 and neglects both cohesion and pore overpressure when calculating the slip tendency of faults. It should be noted that the abbreviation ST is used in the following discussion for both slip tendency in general and for the parameter calculated after Equation 5.7.

5.3. Previous studies on fault reactivation

The ST method has been developed for the analysis of fault reactivation in the southwestern US (Yucca Mountains, Nevada; Morris et al., 1996; Ferrill et al., 1999). (Lisle and Srivastava, 2004) tested the application of this method. They compared the orientations and slip lineations of a large number of natural examples of reactivated faults with the orientations predicted from the ST method. For natural reactivated faults, the orientation of slip is commonly determined by methods of stress inversion that are based on the Wallace-Bott assumption (after Wallace, 1951; Bott, 1959). This assumption states that slip occurs in the direction parallel to the direction of resolved shear stress on a pre-existing fault surface. It implies that the slip direction depends on the character of the stress tensor and the orientation of the fault. The comparison of the ST and the inversion method showed a good match between the orientations of theoretically predicted and natural reactivated fault surfaces (Lisle and Srivastava, 2004). Due to the consistency in results, the authors concluded that the controlling factors for fault reactivation are the magnitudes of fault surface friction (μ) and the available shear stress. Furthermore, Lisle and Srivastava (2004) support the use of ST as a useful parameter for fault instability and earthquake prediction.

In Central European studies, the ST method has been applied to faults in the Lower Rhine Graben and the North German Basin. Worum et al. (2004) have calculated ST for faults of the Roer Valley Graben in the southern Netherlands, which is a seismically active part of the Lower Rhine Graben (for details see Worum, 2004). Using a script for the commercial software gOcad, a 3D stress state was calculated at every location of the fault model. With the stress tensor and the 3D geometries of the faults as input data, ST was calculated on the 3D fault surfaces. For the calculation of the stress field, the orientation of the maximum horizontal stress component (S_H) and the ratios of the three stress components (S_H, S_h, S_V) were used. Data on absolute stress magnitudes for the Roer Valley Graben are not available. Therefore, Worum et al. (2004) used information on the ratios of the stress components, which has been obtained from inversion of focal mechanisms of the Roermond 1992 earthquake sequence (data of Camelbeeck et al., 1994). For the stress field calculations, Worum et al. (2004) assumed a laterally homogeneous stress field that was independent of material properties. Earthquake focal mechanism studies of the area are inconclusive as to whether the stress regime in the Roer Valley Graben is extensional or strike-slip. Therefore, Worum et al. (2004) calculated ST for both stress regimes. The results of ST calculations show that an extensional stress regime is more likely for the area. The authors assume that increased ST occurs on fault segments that have been frequently reactivated during Quaternary times. Therefore, the distribution of ST is compared and directly correlated to the vertical throw values of the main faults in the study area. One key result of this correlation is that the fault reactivation modeled could only match the observed faulting activity when relatively low frictional coefficients were used (μ_{static} of 0.3 – 0.4).

Leydecker (2005b) used the ST method to assess the potential of fault reactivation and the likelihood of damaging earthquakes for the North German Basin, an area of very low seismic activity (compare Figure 1.8). FE modeling was used for the simulation of a 3D stress state that was calibrated with in-situ stress indicator data obtained from boreholes (Connolly et al., 2003). On the basis of the stress field modeled, ST was calculated for several large faults in the study area (Connolly et al., 2003). Leydecker (2005b) used the ST results for further seismological investigations. It was assumed that increased ST values on fault segments indicate the location of potential, future earthquakes. In order to assess the size of potential earthquakes, the measured lengths of continuous fault segments with increased ST have been taken for estimates of possible earthquake magnitudes. The ST values modeled are relatively low. This is interpreted that under the present-day stress field, faults are in a stable state and not prone to reactivation. This conclusion is based on using a coefficient of friction μ of 0.6 (Connolly et al., 2003; Leydecker, 2005b).

Recently, several numerical modeling studies in the URG area have provided insights into the reactivation of intra-graben and border faults. Lopes Cardozo (2004) investigated the displacement pattern of faults in the southernmost part of the URG under present-day stress field conditions. A simple 3D model setup with linear elastic rheology was used with displacement boundary conditions applied over a time span of 1000 years. Gravitational body forces were neglected. The model covered the upper crust from 2.6 to 7 km depth for an area of 50 x 50 km and included 8 fault surfaces. It was not documented how the friction on the fault surfaces was defined. Three different stress field scenarios were modeled using varying magnitudes for the displacement boundary conditions. In the three cases modeled, all faults implemented were reactivated whereas deformation was mainly concentrated on the Eastern Border Fault (EBF) in the vicinity of Basel.

Schwarz and Henk (2005) developed a more complex model setup using 3D thermomechanical modeling with the FE code ANSYS (for details see Schwarz, 2006). This study focused on the structural and depositional evolution of the entire URG during the Tertiary. The model covered a 200 x 275 km area with the URG located in the center of the model. It included the crust and the upper mantle down to a depth of 40 km and the border faults implemented as contact surfaces. The depositional pattern was reproduced by implementation of listric geometries of the border faults. One of the key results of this study was that the depositional pattern was matched only when low friction was assumed for the border faults (μ 0.3 – 0.4).

A 3D mechanical modeling study of Buchmann and Connolly (2007) investigated the contemporary stress state of the entire URG. For this study, the FE code ABAQUS was used. The focus of the study of Buchmann and Connolly (2007) was placed on simulating the present-day state of stress of the URG. The model setup of this regional-scale model was relatively simple and comparable to the setup of Schwarz and Henk (2005), including two crustal layers, the Moho topography and the border faults implemented as contact surfaces. The model covered an area of 500 x 500 km and a depth of 60 km. The simulation of the present-day stress state showed that in contrast to the findings by Schwarz and Henk (2005) reactivation of the border faults at geologically and geodetically observed rates can be explained with friction being slightly higher on the faults (μ 0.6).

5.4. Modeling approach of 3D stress field modeling

The focus of this FE modeling study is to investigate the reactivation potential of faults in the URG and adjacent shoulder areas. The approach chosen resembles the one of Connolly et al. (2003) and Leydecker (2005b). This type of modeling consists of two models: 1) a regional 3D mechanical model for simulation of a 3D stress field of the area of interest, and 2) a 3D fault model comprising the fault surfaces of the area. For this thesis, the 3D stress field was simulated for the entire URG and surrounding shoulder areas. The simulation was carried out with the FE code ABAQUS^{TM}. The 3D stress state was then used for the calculation of the slip tendency parameter on fault geometries using the GeoMoVie post-processing software.

Since slip tendency calculations strongly depend on the input data from the 3D stress field modeled, emphasis in this work was placed on modeling the best-fit and most realistic stress field of the URG area. Initially, a relatively simple stress field model was chosen (Model 1). However, evaluation of the slip tendency results based on this model revealed that slip tendency was overestimated. In addition, the stress field obtained at greater depths was not realistic and not in agreement with independent observation (see discussion in section 5.4.2.2). Therefore, a more complex model was developed in order to obtain a more realistic 3D stress field model (Model 2). This best-fit stress field model was then used for the slip tendency analysis. In order to demonstrate the advantages of the more complex model, both stress field models are presented and discussed (section 5.5).

Both stress field models consider purely elastic and isotropic material behavior. The use of elastic rheology is an oversimplification. However, an elastic rheology is a justifiable approach for investigating the first-order tectonic stresses of the URG area, in particular when the data available is insufficient to constrain a more complex rheology. Stress data available for this area includes local and shallow data on the absolute stress magnitudes at the Soultz-sous-Forêt drilling site and regional data on the stress regimes in the upper crust (Rummel and Baumgärtner, 1991; Delouis et al., 1993; Plenefisch and Bonjer, 1997; Klee and Rummel, 1999; Hinzen, 2003; Lopes Cardozo and Behrmann, 2006). Given this database, the main objective of this modeling study is to obtain a stress field that represents the first-order stress pattern of the URG area. Moreover, the study aims to obtain realistic stress regimes and thus realistic relative stress magnitudes, rather than estimating the absolute stress magnitudes, which are unknown for larger depths anyway. It should be noted that both stress field models do not include fault surfaces. Therefore, this modeling approach does not address effects such as stress release due to fault movement (differential motions of fault blocks). This further implies that strain rates of the model cannot be compared to real data.

In order to simulate a 3D stress field as realistic as possible the loading of the FE model and the definition of the boundary conditions are most crucial. Both 3D stress field models have been loaded with gravity since gravitational forces significantly contribute to the stress tensor by primarily determining the magnitude of the vertical stress component S_V. For the loading of the models from the sides, to simulate the horizontal stress components (S_H and S_h), displacement boundary conditions have been applied. Displacement boundary conditions are commonly used in 3D FE modeling and preferred to the application of forces or pressures since they are technically easier to apply. In the following, both stress field models are described in more detail. A detailed description of the loading of both models is given in section 5.4.2.

5.4.1. Geometries of the 3D stress field models

The geometries of both models are very similar. In addition, both of them are constructed with tetrahedral elements (C3D4 element type in ABAQUS). The area of interest, the URG, is located in the center of the stress field models and is constructed as a circular body including the surface topography (Fig. 5.5). The circular shape has been chosen for its simple geometry. It also covers the entire area of the 3D fault model. A load frame onto which the lateral boundary conditions are applied surrounds the URG volume. Of particular interest to this study are the upper crust and the stresses within it that influence the reactivation potential of the faults in the area. Therefore, both models represent the upper crust of the URG area.

Model 1

For model 1, the circular body is constructed with elements of ~3 km size (~3 km resolution). The load frame has a coarser mesh with ~5 km large elements. The thickness of model 1 is 15 km, representing the upper crust (Fig. 5.5; see also Fig. 5.3).

Model 2

In order to honor the different material properties between the highly faulted sedimentary graben fill and the less faulted basement rock of the shoulder areas, model 2 implements a graben in the center of the circular body (Fig. 5.5). The eastern and western contacts between this graben and the surrounding shoulder areas have geometries corresponding to the eastern and western border faults, respectively. The basement underneath the graben is more faulted than the basement of the shoulder area (Figs. 5.3, 5.4). Therefore, similar material properties have been assigned to the faulted basement underneath the graben and the graben fill. This means for the geometry of model 2 that the graben has been constructed with a maximum depth of 10 km, in contrast to the actual depth of its sedimentary fill of max. 3 km. The depth of 10 km corresponds approximately with the lower extent of basement faults underneath the graben (Fig. 5.3). The graben is constructed with an element size of ~2 km. The shoulder area is constructed with ~3 km size elements and has a depth of 25 km. Model 2 cover the entire crust to a depth of 25 km. The larger depth of model 2 has been chosen in order to avoid boundary effects from the bottom of the model. For ST calculations, stress data of the upper 10 km of the crust is used. A load frame with a mesh of ~3 – 5 km size elements surrounds the internal model.

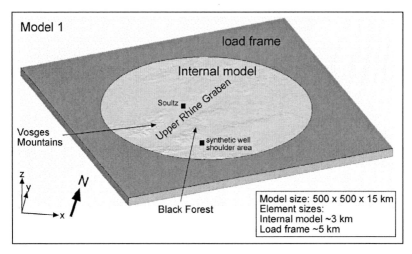

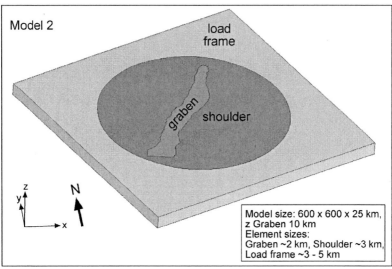

Figure 5.5. Geometries of 3D stress field models. Model 1: the internal model includes the surface topography taken from GTOPO30 database. Note the locations of wells: reference well at Soultz-sous-Forêt in the central URG and a synthetic well in the shoulder area used for comparison of stress data (see also Figure 5.2). Model 2: the internal model includes the same topography as model 1. Additionally, a distinction between graben and shoulder lithospheric material properties is taken into account (see Table 5.2).

5.4.2. Loading of the 3D stress field models

5.4.2.1. Model 1

Model 1 was loaded in two steps (Fig. 5.6). In the first step, the model was loaded with gravitational forces (gravitational pre-stressing step). This initial state of stress was then implemented as initial conditions to an undeformed FE mesh during step 2. The application of both gravity and displacement boundary conditions in step 2 simulated the contemporary tectonic loading of the geological system. The displacement boundary condition was oriented in the direction of the observed average S_H at N150°. In order to simplify the loading with lateral displacements, the boundaries of the model were oriented perpendicular to the displacements (Fig. 5.6). A displacement was applied on the southeastern boundary of the model, whilst the other three boundaries and the bottom of the model were constrained with rollers that allow displacements of nodes in plane of the sides only. During both steps, the top surface was not constrained and thus acted as a free surface (Fig. 5.6).

Other user-defined boundary conditions included the elastic material properties. The model simulates a 15 km thick part of the upper crust as being isotropic and of linear elastic behavior. The user-defined values for the material properties (density, Young's modulus, Poisson's ratio) are "typical" values and have been taken from literature (e.g. Turcotte and Schubert, 2002) and are in the range typically used in numerical modeling of the Rhine Graben system (e.g. Dirkzwager, 2002; Schwarz and Henk, 2005; Buchmann and Connolly, 2007; Table 5.1). Model 1 assumes homogeneous elastic material for the upper crust. Neglecting variations in material properties and using simplified boundary conditions results in a uniform orientation of the stress field. This implies that local variations of the stress field induced by local heterogeneities cannot be addressed with this modeling approach (compare Figure 5.2). On the other hand, model 1 includes a contribution of the topography to the loading.

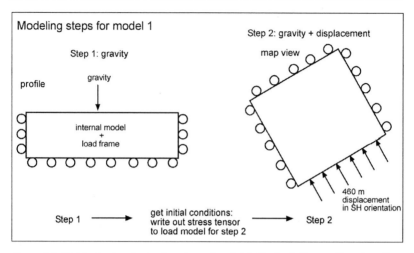

Figure 5.6. Sketch showing the loading steps of model 1. Circles indicate rollers that allow displacements of elements in plane of the sides only. Arrows show the direction and magnitude of lateral displacement applied.

Model 1: Model unit	Rock type	Density ρ [kgm^{-3}]	Young's Modulus E [GPa]	Poisson's ratio ν
Internal model with graben & shoulder	Sediments, granite, sandstone, gneiss	2450	50	0.25

Table 5.1: Mechanical material properties for model 1. For this model, average values are used (see Turcotte and Schubert, 2002).

In order to obtain realistic stress magnitudes, in-situ stress estimates from the geothermal drilling site Soultz-sous-Forêt in the central URG were used for calibration. Figure 5.7A displays the magnitudes of the three stress components and shows the differences in interpretation of the wellbore data obtained from the drilling site (for a detailed review of stress results from Soultz see Cornet et al., 2007). For calibration of the model, the stress estimates of Klee and Rummel (1999) have been used. This has been an arbitrary choice amongst the relatively similar results of Rummel and Baumgärtner (1991) and Klee and Rummel (1999). In order to match the observed stress magnitudes at the Soultz well at a reference depth of 3 km with the stress field model (Fig. 5.7B), a sensitivity analysis of the displacement boundary conditions was conducted. A good fit was reached by applying 460 m of displacement in the direction of S_H (N150°) on the southeastern side of the model.

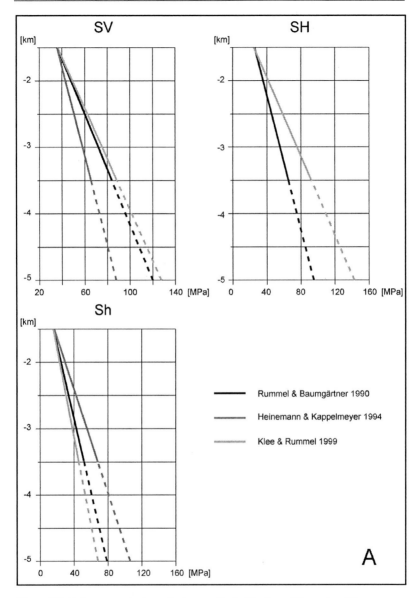

Figure 5.7. A) Stress magnitudes at Soultz-sous-Forêt. The data of Rummel and Baumgärtner (1991), Heinemann and Kappelmeyer (1994) and Klee and Rummel (1999) resulted from well depths of 1.5 to 3.5 km (depth relative to surface). Interpolated data (dashed lines) show how the results vary with depth. For the S_H-curve, only the data of Klee and Rummel (1999) is shown. It is identical with the curve of Heinemann and Kappelmeyer (1994).

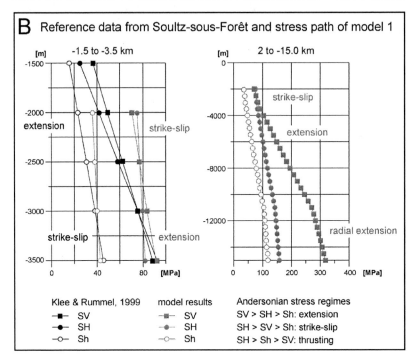

Figure 5.7. B) Stress path at the reference well Soultz-sous-Forêt at -1.5 to -3.5 km (left panel) and down to -15km (right panel). In black stress data of Klee and Rummel (1999), in grey modeling results of model 1. Due to the element resolution, the first modeling data is at 2 km below the surface. Note the changes in tectonic stress regimes, according to Andersonian definition, along the well paths. Model 1 has a nearly good fit with observed data for the upper kilometers. However, with increasing depth this model predicts too much extension compared to observed data.

5.4.2.2.   Problem with the loading of model 1

Comparison of the stress paths modeled and observed shows that the results of model 1 are in agreement with the observed data only around the reference depth of 3 km (Fig. 5.7B). Above the reference depth, the magnitudes of the stress components modeled are slightly higher than the observed values. Below the reference depth, the horizontal stress components modeled are significantly underestimated (Fig. 5.7B). With increasing depth, the differential stress (between the horizontal and the vertical stress components) increases. This results in extremely high differential stresses and a stress state of radial extension at greater depths (below 10 km; Fig. 5.7B). However, this state of stress is not in agreement with observation. Seismological data suggest that the state of stress in the upper crust of the URG area down to a depth of approx. 20 km is transitional between extension and strike-slip (Delouis et al., 1993; Plenefisch and Bonjer, 1997; Hinzen, 2003). The prediction of radial extension at greater depths is not a realistic stress state for the study area. The reason for the unrealistic results of model 1 is primarily the underestimation of the horizontal stresses when loading this model with lateral boundary loads. With increasing depth, the effect of the gravitational force dominates in model 1 and the vertical stress component S_V becomes too large with respect to the horizontal stress components (right panel in Fig. 5.7B).

The loading of model 1 follows a commonly used approach for simulation of a gravity dependent state of stress of an elastic material (e.g. Dirkzwager, 2002; Connolly et al., 2003; Cornu and Bertrand, 2005; Schwarz and Henk, 2005). Herein, a 3D rectangular box with typical rheological parameters of crustal rocks is used to represent a part of the Earth's crust (Fig. 5.6). The boundary conditions on the sides and the bottom are generally defined with rollers (in plane displacement). Loading of a numerical model in the vertical direction only yields magnitudes of vertical stress (S_V) that are a function of the gravitational constant, the depth of the model and material densities assigned. The resulting horizontal stress in this type of model depends exclusively on the vertical stress and the Poisson's ratio assigned. For an isotropic elastic material with constant elastic properties over time, the stress induced in the horizontal directions (S_{Hmean}), is related to the vertical stress (S_V) through the Poisson's ratio (v) by:

$$S_{Hmean} = S_V \left(\frac{v}{1-v} \right) \qquad Eq.\ 5.11$$

The ratio of the horizontal stress magnitude to the vertical stress magnitude is the k ratio (coefficient of lateral stress):

$$k_{mean} = \frac{S_{Hmean}}{S_V} = \left(\frac{v}{1-v} \right) \qquad Eq.\ 5.12$$

On the basis of this relationship, the horizontal stress magnitudes are 1/3 of the vertical stress magnitudes when a Poisson's ratio "typical" for most crustal rocks is assigned (0.2 to 0.25). The k ratio for the crust is ~0.3 consistently with depth (Fig. 5.8). This result is not in accordance with in-situ stress determinations from boreholes. Worldwide observation from in-situ stress measurements indicate that k_{mean} is larger than 1 for the upper kilometer of the crust and between 0.5 – 1.0 for depths of 1 to 3 km (Rummel et al., 1986; Sheorey, 1994). The data points to a compressional state of stress in the uppermost parts of the crust. Stress estimates from the KTB deep borehole support this observation. Values for k_{mean} are in the range of 1.0 in the 9 km deep borehole (Brudy et al., 1997; Fig. 5.8).

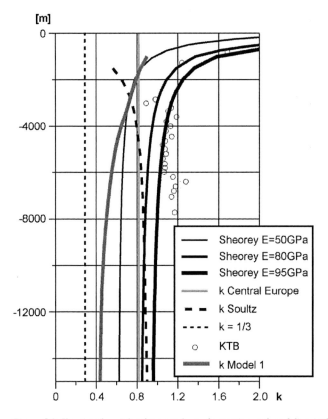

Figure 5.8. K ratios (k_{mean}) for the crust from observation and modeling including model 1. K ratios after Sheorey (1994) are obtained from Equation 5.16, for Central Europe after Equations 5.13 and 5.14., for Soultz after Klee and Rummel (1999) and for KTB site with courtesy provided by T. Hergert (after Brudy et al. 1997). The k ratio of 0.3 results from modeling with typical Poisson's ratios.

Rummel et al. (1986) investigated hydrofracturing stress data from numerous boreholes in Central Europe and determined k ratios for the continental crust of the region:

$$k_h = \frac{S_h}{S_V} = \frac{0.15}{H} + 0.65 \qquad\qquad Eq.\ 5.13$$

$$k_H = \frac{S_H}{S_V} = \frac{0.27}{H} + 0.98 \qquad\qquad Eq.\ 5.14$$

where H is the depth in m. Based on these equations, k_{mean} of Central Europe has a value of 0.82, which implies strike-slip to extensional tectonic regimes for depths where seismicity typically occurs in this area (Rummel et al., 1986; Fig. 5.8). In graben settings, k ratios show the opposite effect: they are relatively low near the surface, indicating extensional tectonics, and increase to a constant value with increasing depth (Klee and Rummel, 1999). This scenario is observed for the URG in the well at Soultz-sous-Forêt (Klee and Rummel, 1999; Fig. 5.8). (Rummel et al., 1986) and Sheorey (1994) conclude that these changes of the k ratio are caused by tectonic activity, major geological structures, topography and heterogeneities in the crust.

Sheorey (1994) proposed the following expression to estimate the k ratio of the Earth:

$$k_{mean} = \frac{v}{1-v} + \frac{\beta EG}{(1-v)\gamma}\left(1 + \frac{1000}{H}\right) \qquad\qquad Eq.\ 5.15$$

where β ($^{\circ}C^{-1}$) is the coefficient of linear thermal expansion, E is the Young's modulus (MPa), G ($^{\circ}C/m$) is the geothermal gradient, γ is the unit rock pressure (MPa/m) and H (m) is the depth. This function was obtained from an elasto-static thermal stress model representing the Earth as a 1D layered and spherical model ranging from the surface to the base level of compaction, the core-mantle boundary. For the crust and mantle, elastic properties, density, temperature gradient and the temperature dependent thermal expansion coefficient of the crust and mantle were considered. The k ratio for the crust, which is the top slice in Sheorey's model, can be given by the simplified equation:

$$k_{meancrust} = 0.25 + 7E\left(0.001 + \frac{1}{H}\right)$$ *Eq. 5.16*

where E in contrast to Equation 5.15 is now in GPa. Sheorey (1994) demonstrated from his analysis that the k ratio results from equilibrium between gravitational compaction and thermal expansion. This implies that crustal stress depends on material properties and variables down to the Earth's core. In addition, the k ratio in the crust depends on the spherical geometry of the Earth.

Figure 5.8 compares the distribution of crustal k ratios from observation and spherical modeling of Sheorey (1994). The figure highlights that the results of model 1 do not match the observed and modeled data. Model 1 significantly underestimates the horizontal stress, in particular for larger depths (4 – 15 km) at which earthquakes and thus fault reactivation typically occurs. This demonstrates that the horizontal stresses generated only by gravity induced elastic compaction are not sufficient to explain the observed crustal stresses.

5.4.2.3. Model 2

In this section, it is demonstrated that underestimating horizontal stresses at depth can be overcome by defining an appropriate initial loading condition for an elastic model. For numerical modeling of the URG stress field Buchmann and Connolly (2007) developed a technique of pre-stressing that enabled simulation of a realistic stress field with realistic k ratios. A more thorough description of the technique and its general application is provided in Hergert et al. (submitted). The concept of this technique has been used for the loading of the improved model 2.

Sheorey (1994) showed that the k ratio in the crust depends on the spherical geometry of the Earth. His 1D model ranged from the surface to the base level of compaction, the core-mantle boundary. In order to avoid using such a large numerical model for the 3D simulation of the URG, a simplified approach was chosen. Herein, a box shape setup similar to model 1 was also used for model 2 representing the upper crust (Fig. 5.5). The stress loading of model 2 was also performed in two steps (Fig. 5.9). In the first modeling step, the model was loaded with gravitational forces (gravitational pre-stressing step), and in the second step, displacement boundary conditions and gravity were applied (tectonic loading step).

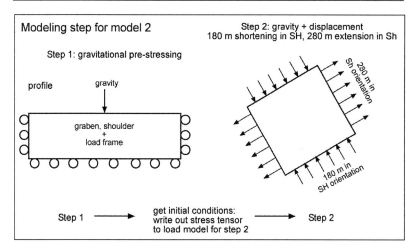

Figure 5.9. Sketch showing the loading steps of model 2. Circles indicate rollers that allow displacements of elements in plane of the sides only. Arrows show the direction and magnitude of lateral displacement applied.

1. Loading step: gravitational pre-stressing

The results from model 1 showed that the Poisson's effect based on typical Poisson's ratios for crustal rocks was insufficient for the generation of horizontal stresses. In order to compensate the effects of compaction a Poisson's ratio higher than the typical value of 0.2 was used for model 2 during the first loading step. The use of a higher Poisson's ratio was necessary to increase the horizontal stress magnitudes within the model in order to obtain realistic k ratios. The variation of the Poisson's ratio accounts for the compaction history of rock volumes. During formation most rocks exhibit a high fluid content both in case of metamorphic or igneous rocks (i.e. during the phase of (re-)crystallization) and for sediments (i.e. water loss during compaction). In this stage, any rock volume has a higher effective Poisson's ratio. During the compaction process over millions of years, the Poisson's ratio reduces to the observed values of 0.25 – 0.3 (e.g. Gercek, 2007). An analogue to the geological compaction history is the compaction, hydration and hardening of concrete where the Poisson's ratio decreases with age (Tsulukidze et al., 1973). Thus the higher Poisson's ratios used in the procedure of simulating the initial state of stress aims to mimic the entire compaction history within a single modeling step.

For the gravitational pre-stressing of model 2, a gradient of Poisson's ratios has been applied (Table 5.2). The necessary Poisson's ratio for specific depths have been obtained through the following relationship:

$$v = \frac{k}{1+k}$$ *Eq. 5.17*

Model 2: Model unit	Rock type	Density ρ [kgm^{-3}][2)]	Young's Modulus E [GPa][2)]	Poisson's ratio v	
				Modeling step 1: Gravitational pre-stressing[1)]	Modeling step 2: Tectonic loading[2)]
Graben (0 – 10 km)	Sediments	2300 - 2700	20 - 61	0.27 - 0.47	0.24 - 0.3
Shoulder (0 – 25 km)	Granite, sandstone, gneiss	2550 - 2790	40 - 61	0.45 - 0.49	0.24 - 0.3

Table 5.2: Mechanical material properties for model 2. References for the materials of model 2 included: 1) v for graben: Klee and Rummel (1999), v for shoulder: Rummel et al. (1986); 2) Turcotte and Schubert (2002)

In this work, two different k paths are used. For the shoulder area the path of Rummel et al. (1986) is used, which is characteristic for basement rocks of the Central European intra-continental setting (Equations 5.13 – 5.14). For the graben fill, the younger compaction history needs to be honored. From the data of the well at Soultz-sous-Forêt, Klee and Rummel (1999) derived formulas for the three stress components at this site:

$$S_h = 15.7 + 0.0149 * (H - 1458)$$ *Eq. 5.18*
$$S_H = 23.5 + 0.0337 * (H - 1458)$$ *Eq. 5.19*
$$S_V = 33.1 + 0.0264 * (H - 1377)$$ *Eq. 5.20*

These stress relations are used to obtain the appropriate k ratio and subsequently the values for v necessary for the graben fill. Table 5.2 summarizes the elastic material properties for model 2.

Figure 5.10 shows the k paths for the graben and the shoulder after the gravitational pre-stressing step. The gravitational pre-stressing results in a model with the vertical stress magnitudes, S_V, being in the right order in magnitude. In addition, the resulting ratio of horizontal to vertical stresses, the k ratio, fits the available calibration data for crustal depths below 3 km. For shallower depths, the model slightly deviates from the calibration data. In the shoulder area at shallow depths, the state of stress modeled is less compressive than predicted by Sheorey (1994) and in the graben the stress is not as extensional as predicted by Klee and Rummel (1999). The less compressive stresses in the shoulder area result from neglecting the spherical shape of the Earth. However, keeping in mind that this stress field model is used to calculate the reactivation potential of faults that generally reactivate at depths below 3 km (range of focal depths in the URG area ~3 – 12 km), the modeling results satisfy the requirements because they deliver a good match for the depth level of interest.

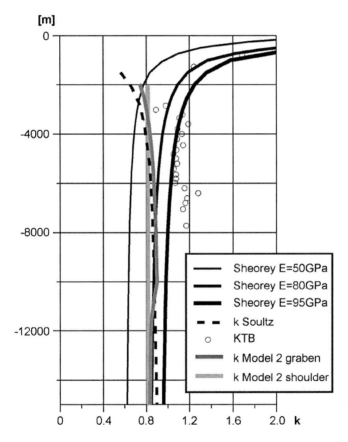

Figure 5.10. K ratios for the crust from observation and modeling supplemented with results from model 2 from the graben (location of Soultz-sous-Forêt) and the shoulder area. The k ratio predicted by model 2 is calibrated in the graben using data of Klee and Rummel (1999), who suggest extensional tectonics in the upper kilometers. The k ratio for the shoulder area is calibrated with data of Rummel et al. (1986) for Central Europe. The curve of Rummel et al.

(1986) is not shown since it is identical with the curve modeled. The k path of Sheorey (1994), accounting for general crustal conditions, predicts slightly higher compression near the surface than the data of Rummel et al. (1986) and the results modeled.

2. Loading step: tectonic loading

In order to obtain a model that also exhibits the horizontal stresses in the correct order in magnitude, the model is loaded in a second step by lateral displacements (Fig. 5.9). The initial state of stress from step 1 was implemented as initial conditions to the undeformed FE mesh for the loading step. For this initial stress state and during the tectonic loading step, Poisson's ratios for compacted material ("typical" values), representing present-day material properties, are assigned to the modeled rock volume (Table 5.2). It should be noted that the change to lower Poisson's ratios to the initial conditions has no effect on the stress magnitudes or the stress field distribution. When only the Poisson's ratios are changed during the initial conditions, the magnitudes of the vertical stresses do not change owing to the free upper surface. Consequently, the horizontal stresses do not change either. The horizontal stress magnitudes change only when lateral displacements are applied on the model boundaries during the second loading step (tectonic loading). During this step, the displacement boundary conditions have been changed such that the simulated stress magnitudes at the Soultz well at 3 km depth fit the in-situ data of Klee and Rummel (1999; Fig. 5.11). This was reached by shortening of the model in the direction of S_H by 180 m and by extension in the direction of S_h by 280 m.

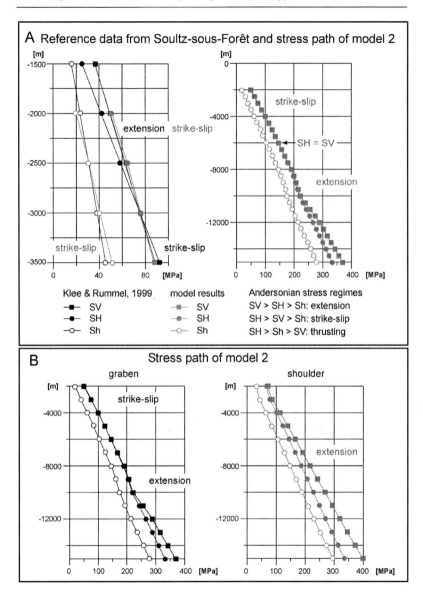

Figure 5.11. A) Stress path at the Soultz reference well from -1.5 to -3.5 km (left panel) and down to -15km (right panel). In black stress data of Klee and Rummel (1999) from the well of Soultz, in grey modeling results of model 2. Due to the element resolution, the first modeling data is at 2 km below the surface. For the upper 2 – 3.5 km, model 2 has a much better fit with the reference data than model 1 (compare Fig. 5.7B). Note the changes in tectonic stress

regimes from strike-slip to extension along the well paths. These regimes are in agreement with earthquake data. B) Stress path of model 2 for the graben and shoulder. For the upper kilometers, the graben is slightly more compressive than the shoulder area.

5.5.   Results of 3D stress field modeling

5.5.1. Stress regime

Both stress field models have been calibrated with in-situ stress data from the Soultz-sous-Forêt well in the graben interior. From the distribution of stress magnitudes with depth, the Andersonian stress regimes can be defined (Figs. 5.7B and 5.11). The differences between the two models become obvious when comparing the stress paths at greater depths. Model 1 shows relatively large stress regime variations ranging from strike-slip near the surface to radial extension below 10 km depth (Fig. 5.7B). In contrast, model 2 shows a good fit with the observed data in the well at the depth range of 2.0 to 3.5 km (Fig. 5.11A). The reference data shows that the magnitudes of S_H and S_V are nearly equal, in particular at the reference depth of 3 km. This explains the transitional state of stress at Soultz-sous-Forêt between extension ($S_V > S_H > S_h$) and strike-slip ($S_H > S_V > S_h$) that is also characteristic for the entire URG area (Plenefisch and Bonjer, 1997). Since the model is calibrated to the data of Klee and Rummel (1999) at Soultz, S_H and S_V are also close in magnitudes. Due to slight differences in the magnitudes of S_H and S_V between the reference data and the model, different stress regimes are obtained for the upper 3 km; extension for the reference data and strike-slip for the model. Below 6 km, extension is the dominant tectonic regime in the model 2 graben. In this model, the shoulder areas are from the near surface to the depth of 15 km under extension (Fig. 5.11B). To conclude, the stress path obtained for model 2 seems more realistic than the stress path of model 1. Furthermore, model 2 is in close agreement with seismological and geological observations pointing to a transitional stress regime between strike-slip and extension and/or transtension for the study area (Plenefisch and Bonjer, 1997; Lopes Cardozo and Behrmann, 2006).

In order to display in addition to the Andersonian stress regimes also transtension and transpression from the modeling results, the parameter referred to as the Regime Stress Ratio (RSR) is used. The RSR value is a combination of the Andersonian stress regime (Anderson, 1905; Anderson, 1951) and the stress ratio after Bott (1959). The stress ratio describes the ratio of the principal stress magnitudes and ranges from 0 to 1:

$$R = \frac{\sigma_2 - \sigma_3}{\sigma_1 - \sigma_3}$$

Eq. 5.21

Values of 0, 1 and 2 are assigned for extension, strike-slip and thrusting stress regimes, respectively. The RSR value is then defined as (e.g. Simpson, 1997):

$$RSR = (\text{regime} + 0.5) + (-1)^{\text{regime}}(R - 0.5)$$

Eq. 5.22

RSR values describe a continuous range of deformation style from radial extension to compression, whereas 0 is radial extension and 3 is compression. Transtension is represented by an RSR value of 1. It should be noted that RSR is equivalent to the Stress Index R' of Delvaux et al. (1997).

Figure 5.12A shows RSR values along the well of Soultz for both stress field models. In comparison to the Andersonian stress regimes displayed along the well (Fig. 5.11A, B), the RSR values indicate that model 2 is for most depths and regions characterized by transtension (maps and profiles in Figure 5.12). In map view, minor differences between the graben and the shoulder are predicted, whereas the shoulders are slightly more extensional than the graben. Within the URG, the central graben segment is more compressional than the northern and southern segments (transtension in central graben vs. extension in the North and South).

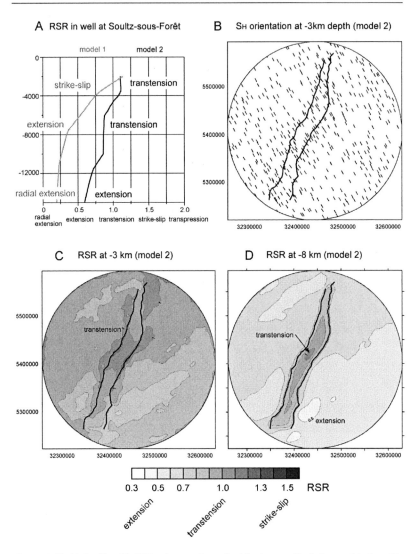

Figure 5.12. A) Profile with RSR values at the well of Soultz-sous-Forêt for models 1 and 2. Model 1 shows strong RSR changes with depth, whereas model 2 is characterized by transtension in most of the upper crust. B) Orientation of S_H for model 2. The general orientation is N150°. Inside and near the graben S_H is rotated counterclockwise by about 5°. C) and D) RSR values for model 2 at 3 and 8 km depths. RSR was calculated using the software GeoMoVie. Coordinates are in UTM coordinate system.

It is interesting to note that the topography implemented in the model has an effect on the predicted stress regime in the upper crust. These topographic effects on the regional stress field and fault reactivation have also been identified in other regions, e.g. for the Western Alps, where large height differences cause stresses of topographic origin (Illies et al., 1981), for the Pannonian Basin that is affected by stresses induced by the surrounding mountain ranges (Bada et al., 2001), or for the San Andreas Fault system in the vicinity of San Gabriel and San Bernadino mountains (Liu and Zoback, 1992). The distribution of RSR values in the shoulder areas for 3 km depth shows that low RSR values are attributed to regions with high topography (Black Forest, Vosges Mountains, Hunsrück Mountains). In these areas, the topography induced extra overburden. This leads to an increase in the magnitude of S_V so that the regime shifts towards extension. The regime change within the URG is related to factors other than topography since the intra-graben surface topography is almost flat. A plausible explanation for the shift to a more compressional regime in the central graben segment is the orientation of S_H with respect to the border faults, which is at a relatively high angle of 70° for this graben segment (Fig. 5.12B). The nearly perpendicular orientation of S_H leads to compression of the soft graben fill in the central segment and an increase of the horizontal stresses. Indications for this more compressive regime have been previously reported by Illies and Greiner (1978), who observed thrusting of the border faults in outcrops. In contrast, the northern and central segments are subjected to shearing under the oblique orientation of S_H with respect to these graben segments.

5.5.2. Stress magnitudes and differential stress

Figure 5.13 displays the absolute stress magnitudes derived from model 2 at 3 km depth. The distribution of S_V and S_h as well as the difference of stresses S_V-S_h reflect the effect of topography on the modeling results. At 3 km, depth the shoulder areas are characterized by extension (compare Figure 5.11B). The comparison of S_V-S_h with the differential stress of the principal stress axes σ_1-σ_3 shows that they are nearly identical, indicating that, at 3 km depth, the principal axes σ_1 and σ_3 are oriented almost vertically and horizontally, respectively. The differential stress $\sigma_D = \sigma_1$-σ_3 is critical for rock failure since an increase in σ_D brings the rock closer to shear failure. The σ_D at which failure occurs is also termed the yield strength of the rock. Figure 5.14 shows σ_D profiles for both stress field models. For comparison, a σ_D profile from the

Soultz-sous-Forêt site, from the KTB deep borehole in southeast Germany and strength envelopes for different faulting styles in the lithosphere are given (calculated after Sibson, 1974). The diagram shows that the differential stress of the two stress field models is relatively similar in magnitude for the upper 2 to 3 km. With increasing depth, the models deliver different results, whereas the differential stress of model 1 increases significantly compared to model 2.

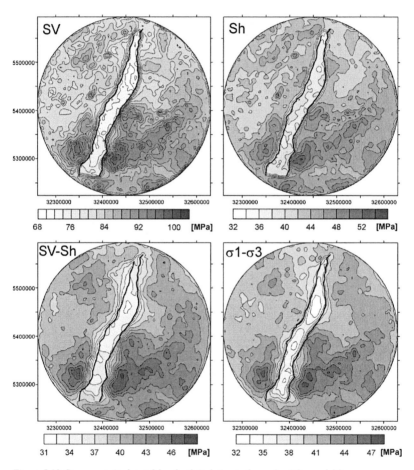

Figure 5.13. Stress magnitudes at 3 km depth (relative to the surface) for model 2.

The σ_D profile of model 2 lies within the field of extensional faulting and shows relatively low strength. This indicates that in the simulated stress field of model 2 only extensional faulting can possibly occur. With increasing depth, the σ_D profile lies further away from the strength profile for extensional faulting and fault reactivation becomes less likely. In contrast, model 1 is critically stressed and pre-existing faults in this stress field simulated would easily be reactivated. Assuming hydrostatic pore pressure conditions, the data at Soultz lies at the boundary between extensional and strike-slip faulting (Fig. 5.14). At the reference depth of 3 km, the differential stress simulated from model 2 has a relatively good fit with the data of Klee and Rummel (1999). However, for larger depths (3.5 to 5 km) the differential stress at Soultz is higher than for the model. It should be noted that for this depth range Klee and Rummel (1999) used extrapolated data (Fig. 5.7A). The authors performed a stability analysis of their data using a simple friction law based on equation 5.8. The results reveal that for a friction coefficient of $\mu = 0.85$ and a fluid pressure slightly higher than hydrostatic pressure, slip on optimally oriented faults is likely. Klee and Rummel (1999) conclude that the upper crust at the site of Soultz-sous-Forêt is just in a stable state under hydrostatic pressure conditions. The σ_D profile from the KTB borehole is based on mean σ_D values after Brudy et al. (1997). The stress regime derived from the borehole is strike-slip (Brudy et al., 1997). In contrast to the URG data, the KTB σ_D profile points to relatively high differential stresses in the upper crust and considerable strength. Assuming hydrostatic conditions, failure of strike-slip faults is likely. In a compilation of European lithospheric strength in form of a 3D strength map, it becomes evident that the KTB is situated in relatively strong crust while the URG as well as the other segments of the European Cenozoic Rift system coincides with zones of weakened lithosphere in the Northwest European Platform (Cloetingh et al., 2005).

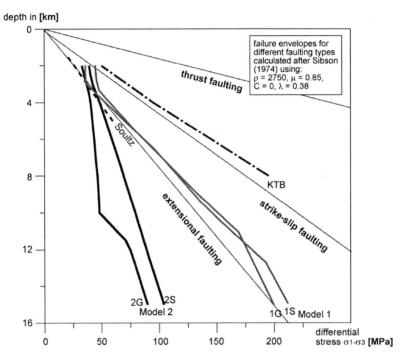

Figure 5.14. Differential stress profile of the modeling results. The differential stress σ_1-σ_3 is given for the reference well in the graben (G) and the shoulder area (S) for model 1 (grey) and model 2 (black). The diagram shows that with increasing depth, σ_1-σ_3 for model 1 is significantly higher than for model 2. The differential stress at Soultz (black dashed line) has been estimated assuming favorably oriented faults with a friction coefficient of μ = 0.85. The estimates are based on the assumption that S_V and S_h correspond to σ_1 and σ_3 respectively (Klee and Rummel, 1999). The σ_D profile of the KTB (black dashed line) is after Brudy et al. (1997) indicating a high-strength upper crust (μ = 0.65). The thin black lines represent strength envelopes for extensional, strike-slip and thrust faulting. The envelopes are calculated after Sibson (1974) for favorably oriented faults and a constant coefficient of static friction μ = 0.85. Other parameters included are a medium density for basement rocks ρ, pore fluid factor λ (ratio of pore fluid pressure to vertical stress) that is here defined to represent hydrostatic pressure (λ = 0.38) and cohesion C defined to be zero.

5.5.3. Summary and comparison of models 1 and 2

The presentation of results from the two numerical models shows that model 1 has only at a depth of 3 km a good fit with the reference data of Soultz-sous-Forêt. With greater depths, the horizontal stresses are underestimated, which results in a stress regime of radial extension and too high differential stresses. This model is too critically stressed. Using model 1 as input for slip tendency calculations would significantly overestimate the slip tendency. The stress field simulated from this model implies that the URG area is highly active with respect to faulting activity / earthquakes. This is not in agreement with the low level of seismicity and faulting observed.

In contrast, model 2 delivers a more realistic stress field scenario. For most crustal depths, this model predicts a transtensional stress regime, which is in agreement with observations (Plenefisch and Bonjer, 1997; Lopes Cardozo and Behrmann, 2006). The differential stress obtained from model 2 is relatively low for depths between 4 and 15 km. This model is less critically stressed than model 1 or the extrapolated data of Soultz-sous-Forêt (Klee and Rummel, 1999). Therefore, this model might underestimate the potential of fault reactivation, particularly at greater depths. However, the results of this model are considered more realistic. Thus, calculations of the slip tendency parameter on the basis of the stress field obtained from model 2 are regarded to deliver more realistic results than those of model 1. Therefore, this work proceeds with model 2 and uses the stress field results from this model as input for the slip tendency calculations.

5.6. Modeling approach for the 3D fault model

5.6.1. Geometry of the 3D fault model

For the construction of a 3D fault model of the URG faults, data from several maps has been compiled (Fig. 5.4). Based on this fault map 3D fault surfaces have been constructed using the commercial software gOcad (Fig. 5.15). The faults with known dip directions are assigned a constant dip of 60° in the 3D fault model, which is a typical dip value for extensional or oblique moving faults. Faults without a known dip direction, mainly located in the shoulder areas, have been assigned with a vertical dip. It should be noted that the dip has significant influence on slip tendency calculations since shear and normal stress acting on fault surface depend on the dip and surface orientation. The vertically dipping faults are not optimally oriented for reactivation in the extensional stress field of the URG area. This implies that slip tendency results for these faults are underestimated. However, the dependency of the fault orientation on the slip tendency remains and identification of fault segments with relatively increased slip tendency is possible (see also section 5.7.1).

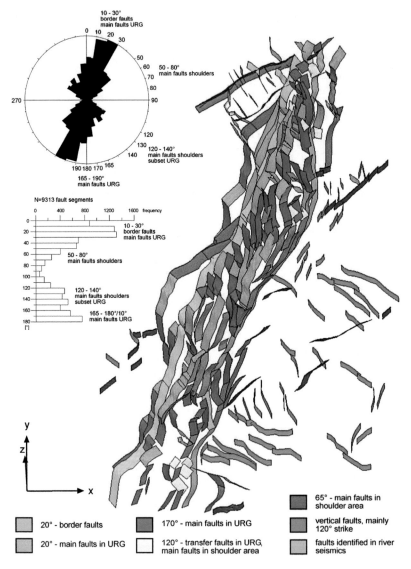

Figure 5.15. 3D fault model discretized with Hypermesh. The compilation of the northern URG faults is based on previous fault mapping (Andres and Schad, 1959; Straub, 1962; Behnke et al., 1967; Breyer and Dohr, 1967; Illies, 1967; Illies, 1974a; Derer, 2003; Derer et al., 2003; Haimberger et al., 2005, and G. Wirsing, pers. comm. 2005). For the southern URG the

mapping of Breyer and Dohr (1967) and Tietze et al. (1979) is used. The rose plot and the histogram show the orientations of all faults displayed, and the grouping into 4 main fault sets. The colors indicate the separation of faults into fault sets based on their orientation and dip.

The definition of the intra-graben faults displayed in Figure 5.4 has mainly resulted from intensive hydrocarbon exploration for Oligocene and Miocene reservoirs at ~ 1 – 2 km depths during the 1960s and 1970s (e.g. by Andres and Schad, 1959; Straub, 1962; Behnke et al., 1967; Breyer and Dohr, 1967). Therefore, for the construction of 3D fault surfaces in gOcad it is assumed that the location of these intra-graben faults on the map represents their location at 2 km depth. For graben-exterior faults, the surface traces are used since these faults have mainly been mapped from rock exposures at the surface. Based on these assumptions, the fault surfaces are constructed to extend from the surface to the typical seismogenic depth of the area at 10 km. The intersection of faults and the definition of antithetic and synthetic faults are based on own interpretations. The fault geometries are then imported to the pre-processing software HYPERMESH that is used for remeshing, construction of a triamesh and model definition for the analysis with the FE code ABAQUS. The element size of the triamesh and thus the resolution of the 3D fault surfaces are defined in HYPERMESH to ~3 km.

5.6.2. Separation of fault sets

In order to discuss the effects of fault orientations on the slip tendency results the faults are grouped into four sets according to their orientation (Fig. 5.15). These fault sets are characterized by the following average strike orientations (in degree): 1.) 020°, 2.) 170°, 3.) 120°, and 4.) 065°. Additional combinations of fault orientations discussed are all vertical faults and a set of faults that have been identified in river seismic profiles of Haimberger et al. (2005) and G. Wirsing (pers. comm., 2005). The 020°-fault set includes the border faults and a large number of relatively long intra-graben faults (main faults of URG). Long intra-graben faults are also a member of the set with 170° orientation (main faults of URG of 170°-fault set). The 120°-fault set includes a few secondary faults of the URG and long faults of the shoulder areas, most of which have been assigned a vertical dip. The 065°-fault set contains mainly long fault segments of the shoulder areas. In the analysis of the slip tendency parameter, the effect of fault orientations on their reactivation potential will be investigated.

Additionally, special attention is paid to the slip tendency results of faults known to be active during the Pleistocene from Figure 5.4, with the aim to correlate between geological evidence for high faulting activity and geomechanical results.

5.7. Slip tendency analysis of Upper Rhine Graben faults

5.7.1. Introduction

In this section, the slip tendency (ST) results are presented and discussed. For ST calculations, the 3D stress field data and the 3D fault surfaces are loaded into GeoMoVie. The software calculates ST using the normal of each triangular facet of the fault mesh using Equation 5.7 and displays the values. Dependent on the resolution of the fault mesh, ST data points are calculated every 2 to 3 km. The results in map view are visualized with Surfer8, with each data point represented by a single symbol in order to demonstrate the resolution. The ST results on fault profiles are interpolated and shown with contours to highlight the spatial distribution of ST.

ST calculations in this study are based on assumptions that:
- all faults are defined to be cohesionless (C = 0), and
- all faults have the same coefficient of static friction μ_{static} (or simply μ).

The choice of μ is crucial since it directly influences the value of ST. Lowering μ increases the ST values and brings the faults closer to reactivation, whereas a higher μ decreases ST and makes fault reactivation less likely. Since the absolute values of μ for URG faults are unknown, ST calculations are performed for various scenarios with μ ranging between 0.2 and 0.6. In addition, the absolute values of the stresses have a large impact on the value of ST. It has been mentioned earlier that model 2 predicts more realistic stress regimes for the URG area. However, the model might underestimate the absolute values of stresses for larger depths, resulting in a too low ST at depth. This shows that for both input parameters required for ST calculations the absolute magnitudes can only be estimated and it can only be assumed that they are in the right order in magnitude. It is thus beyond the scope of this study to predict absolute ST values for the URG area. The aim of this study is to identify those fault segments that are most prone to reactivation – a relative grading between the fault segments is important to consider.

5.7.2. Slip tendency results

5.7.2.1. Slip tendency variations with depth

The stress field model reveals increasing stress magnitudes with depth and changes in tectonic regimes. It is thus expected that ST also varies with depth. The distributions with depth of the input parameters used for ST calculations, shear and normal stress, and the ST values itself are displayed on a profile of the Eastern Border Fault (EBF; Fig. 5.16A). The shear and normal stresses both increase with depth, however the normal stresses increase by much greater degree. This leads to higher ST values near the surface and significantly lower values at typical hypocenter depths (Fig. 5.16A). It has been mentioned earlier that model 2 might underestimate stresses and therefore underestimates ST. This situation becomes evident when plotting the results as Mohr circles (a diagram of normal stress versus shear stress). Figure 5.16B shows Mohr circles at 3 and 8 km depth at the location of Soultz-sous-Forêt. The diameters of the circles are defined by the differential stress at that location. With depth, the differential stress of model 2 increases by a small amount and thus the diameter for the circle increases only slightly (compare Figure 5.14). The upper line represents a failure envelope for $\mu = 0.6$. The model lies in the stable part well below the failure envelopes and fault reactivation on optimally oriented faults is not predicted. Comparing the distribution of shear and normal stresses with depth from Figure 5.16A with the Mohr diagram (Fig. 5.16B), the low shear stresses of model 2 can be explained by the small Mohr circles due to the small differential stress. The increase in normal stress with depth for this model is a result of the increase of both stress components (σ_1 and σ_3) by more or less equal amounts that leads to a shift of the Mohr circle to the right. The Mohr circle delivers also information on the value of μ necessary to cause slip on optimally oriented faults. The occurrence of slip, defined as ST = 1, is only likely for $\mu < 0.3$ (Fig. 5.16B). In absolute terms, model 2 predicts that earthquake activity at typical hypocenter depths is rather unlikely.

It should be noted again that this study aims to identify fault segments with relatively increased ST instead of predicting fault reactivation based on absolute ST values. The profile of the EBF allows distinguishing several segments with increased ST. At all depths, high ST values are located north of Freiburg (FR) and near Heidelberg (HD). For the latter fault segment, recent tectonics and thick Quaternary deposits (*Heidelberger Loch* as the main Quaternary depocenter) are documented (e.g.

Bartz, 1974). In contrast, low ST values are found in the central EBF segment (Fig 5.16A). In the following section, relative ST changes on fault segments of the entire fault set of the URG region are explored.

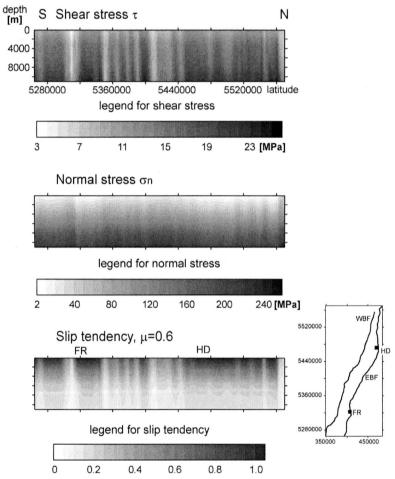

Figure 5.16. A) Shear and normal stresses and slip tendency displayed on profile of the Eastern Border Fault (EBF) using input data from model 2. Normal stresses are relatively high at depth while shear stresses are low. This results in low ST at depth. EBF segments with high ST values are located near Heidelberg (HD) and north of Freiburg (FR), whereas low ST values are found in the central EBF segment. Scale and orientation is the same for all three profiles.

B: Mohr circles from the location of Soultz-sous-Forêt

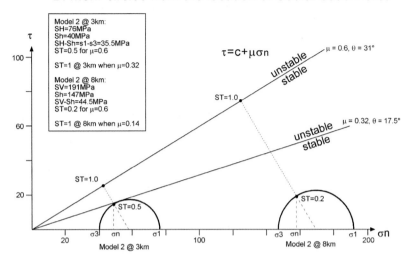

Figure 5.16. B) Mohr circles for model 2 at 3 and 8 km depths at the location of Soultz-sous-Forêt. For a coefficient of friction $\mu = 0.6$ the model is in a stable state (angle of sliding friction $\theta = 31°$). Reactivation of optimally oriented faults at larger depths for the model is only possible with very low friction coefficients ($\mu = 0.14$). The Mohr diagram visualizes that stresses in the model are relatively low and possibly underestimated for fault reactivation to occur.

5.7.2.2. Slip tendency for all faults and individual fault sets

Figure 5.17 gives an overview of the ST results for all faults investigated. In order to demonstrate how the lowering of μ increases the value of ST, the results are shown for μ of 0.4 and 0.6. Here, a decrease of μ of 0.2 roughly increases ST by about 0.2 (Fig. 5.17). In addition, the figure highlights the decrease of ST with depth because of the low differential stresses at depth. The ST results of all faults show that the magnitude of ST strongly depends on the fault orientation. NE-SW trending faults show relatively low ST, while N-S trending faults exhibit the highest ST values (Fig. 5.17).

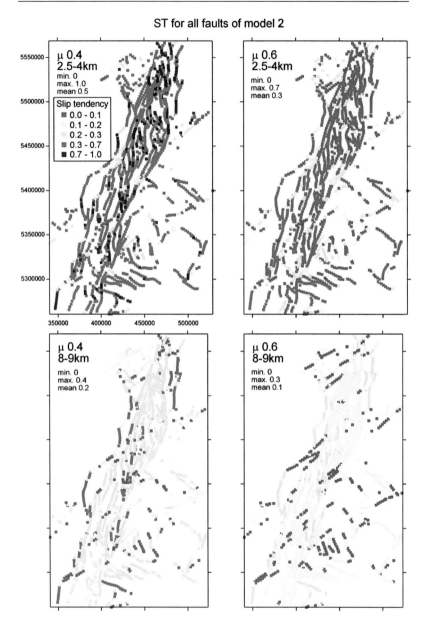

Figure 5.17. Map view of slip tendency of all faults for model 2 at 3 and 8 km depth (relative to the surface) and for μ = 0.4 and μ = 0.6.

To highlight the effect of fault orientation on ST, the results are displayed in Figure 5.18 for the individual fault sets. Faults with the largest ST values strike 170°, which is nearly parallel to the direction of S_H. Note also that the long segment of the EBF in the northern URG (segment of Heidelberg) with the highest ST values has the same strike. In addition, high ST occurs on shorter segments of the EBF and WBF with ~ N-S striking orientations. Faults with intermediate ST values form part of the sets with 020° and 120° strike. The lowest ST values are clearly on faults segments with 065° strike that are oriented nearly perpendicular to S_H. The majority of faults within the URG is favorably oriented and generally exhibits the highest ST values. In contrast, a large number of faults of the shoulder areas are less optimally oriented for fault reactivation.

5.7.2.3. Comparison of slip tendency results with independent studies

Active faults

In the framework of this thesis, active fault segments in the northern URG have been identified by paleoseismological trenching and geomorphological analysis (*Chapters 2 to 4*). In addition, active faults in the entire URG region have been compiled from literature. A map of all active faults is given in Figure 5.4. The ST results for these faults are shown in Figure 5.19. Most active faults strike NNE to SSW and belong to the fault sets of 020° and 170°. From the overview of ST results on all faults, it has been demonstrated that 020° and 170°-fault sets exhibit the highest ST (Fig. 5.18). Note that the fault associated with the Rastatt 1933 earthquake is not favorably oriented for reactivation. Of special interest to this study are active faults in the northern part of the URG. The ST results for this region show several long fault segments with ST values larger than 0.6 (depth = 2.5 – 4km, μ = 0.4). Amongst these segments are: the WBF scarp identified as active from trenching (1), all faults identified in river seismic profiles (2 – 5); a fault segment associated with a terrace scarp morphology (7) and the EBF segment adjacent to the *Heidelberger Loch* (8).

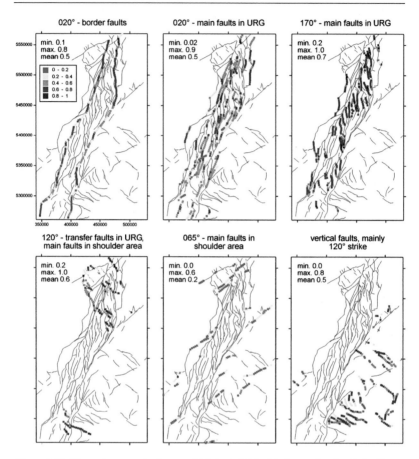

Figure 5.18. Slip tendency results from model 2 for individual fault sets. Depth = 2.5 – 4km, μ = 0.4.

Several faults exhibit a tectonic morphology, which has been identified from the geomorphology analysis of this thesis: the HTBF, the southern border of the Niersteiner Horst, the Pfrimm Valley fault (6 in Figure 5.19), the Riederbach Valley fault and a fault that coincides with the lower terrace scarp (9). It is suggested that these faults have been active during the recent evolution of the northern URG (*Chapters 3 and 4*). Note that the Riederbach Valley fault is not included in Figure 5.19 due to resolution. However, it exhibits the same strike as fault 6 and therefore the ST results are very similar. The faults identified from geomorphology are included in the 065°-fault set and apparently exhibit the lowest ST results of all faults considered (faults marked with green symbols in Figure 5.19). An exception is the fault marked 9 with a 020° strike and intermediate ST values. These results show that in contrast to the geomorphological analysis, the ST analysis suggests that these faults are not prone to reactivation (see discussion in section 5.8.2).

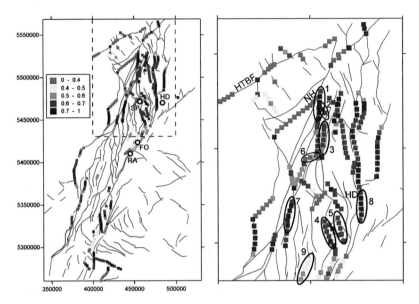

Figure 5.19. Slip tendency on active fault segments. Left panel shows results for entire URG. Right panels shows results of area of interest (dashed box in left panel). RA and FO mark the fault segment related to the Rastatt 1933 and Forchheim 1938 earthquakes respectively. SP and HD show locations of earthquakes near Speyer and Heidelberg from 2005 (M. Wagner, pers. comm. 2007). Faults numbered are of special interest to this study: 1 – trenching (Chapter 2 of this thesis); 2, 3 – river seismic profiles (Haimberger et al., 2005); 4,5 – river seismic profiles (G. Wirsing, pers. comm. 2005), 6, 7, 8, 9 – geomorphology (Chapters 3 and 4). HTBF – Hunsrück-Taunus Boundary Fault. NH – Niersteiner Horst. Slip tendency results displayed for depth = 2.5 – 4km, μ = 0.4.

Orientation of earthquake nodal planes

Focal plane solutions have been inverted for a large number of earthquakes in the URG region, most of which occurred in the southern URG (Plenefisch and Bonjer, 1997). A focal plane solution consists of two nodal planes. One of the planes is created or reactivated during an earthquake. However, it is not possible to distinguish the active plane. Comparison of orientations of nodal planes with those of pre-existing faults in the URG area can provide information on the possible active plane. Figure 5.20 shows the orientations of nodal planes determined by Plenefisch and Bonjer (1997). The authors separated the data set regionally. Only five earthquakes were recorded in the northern URG and adjacent shoulder area, whereas ten times more earthquakes occurred during the same period in the southern URG region (recordings between 1976 – 1992). The northern URG data set includes relatively consistent nodal planes with strike maxima at 0° and 060°. On the contrary, the large data set for the southern URG is less consistent. Several maxima can be distinguished with strike orientations of 0°, 020°, 090° and 100°. When comparing the nodal planes with pre-existing faults of the individual regions, the following observations can be made. Both nodal planes and fault segments in the northern URG have maxima with 0° to 020° strike. The maximum of 060° strike for nodal planes is not represented in the data of mapped fault planes. It is therefore reasonable to suggest that the 0° to 020° nodal planes are the active planes. This observation appears to be supported by the ST results. Fault segments with 0° to 020° orientation exhibit relatively high ST values, whereas the 060°-fault set exhibits the lowest ST of all faults sets distinguished (Fig. 5.18). Note that the 170°-fault set is not present in the nodal plane data. This could be related to the limited number of nodal planes determined, which might not be a representative data set. For the southern URG data, maxima at 0° to 020° exist both for the nodal planes and fault segments. Other maxima do not coincide; in particular, the 090° to 100° strike of nodal planes is not represented in the fault segment population. This could lead to the same interpretation as for the northern URG in that the 0° to 020° nodal planes represent the active planes, whereas the 090° to 100° nodal planes are not active. In addition, the 120° nodal planes could be active since this orientation is characterized by intermediate ST values. Again, the 170°-fault set is not present in the nodal plane data set.

Recently, focal mechanisms of two earthquakes in the northern URG from the year 2005 have been determined (M. Wagner, pers. comm. 2007; Fig. 5.19). For the earthquake near Speyer, a 023° striking nodal plane is suggested (earthquake from 10.02.2005, M_L 2.8). The earthquake near Heidelberg is possibly associated with a 120° striking nodal plane (earthquake from 26.03.2005, M_L 2.4). These results support once more the interpretation that seismicity in the URG preferably occurs on 020° and 120° striking fault surfaces.

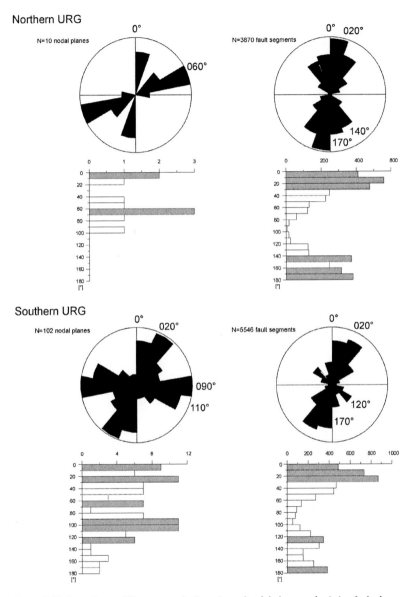

Figure 5.20. Rose plots and histograms of orientations of nodal planes and existing fault planes of the northern and southern URG region. Both regions include the graben and the adjacent graben shoulder. Nodal planes after Plenefisch and Bonjer (1997) from earthquake recordings between 1976 – 1992. The majority of earthquakes have magnitudes of M_L 1 – 3. The few number of nodal planes from the northern URG have maximum strike directions of 0° and

060°. In the southern URG, nodal planes vary in orientation. Maxima are at 020° and 100°. For comparison, fault strikes for both regions are shown.

Tertiary and Quaternary deposition along the border faults

In the following, ST results are compared to fault displacement data of the URG region. Information on displacements on URG faults are available only for few faults and for different periods. Thus, a fault displacement map for a common period, e.g. Tertiary or Quaternary, does not exist. However, information on the thickness of Tertiary and Quaternary graben deposits may be used as a proxy for identifying synsedimentary faulting activity. The main faults in the study area are the graben border faults with estimated vertical displacements of several hundred to thousands of meters (e.g. Pflug, 1982). These estimates are based on the generally accepted concept that large thicknesses of sediments along the faults are related to large vertical, synsedimentary fault offsets. Figure 5.21 shows the thickness of Quaternary deposits along the border faults. The EBF exhibits the largest offset variations. The maximum of 350 m is located adjacent to the *Heidelberger Loch* in the northern URG. A second maximum is found north of Freiburg. The very northern, the central segment and the very southern segments show the lowest offsets of 20 m and less. The WBF has generally low offsets of 20 m. Larger offsets occur at the northern and southern ends of the URG, i.e. north of Oppenheim and south of Colmar.

The profiles in Figure 5.22 show the distribution of Tertiary and Quaternary depocenters and the depth distribution of ST on the border faults. A clear correlation between large sedimentary thicknesses, large vertical offsets and high ST is evident for the *Heidelberger Loch* segment. Along the WBF, this correlation is present for the WBF scarp and the segment near Colmar. At the low activity end of the spectrum, low deposition, low offsets and low ST results coincide along the EBF segment of Rastatt and along the WBF south of Neustadt an der Weinstrasse. Despite this relatively good correlation, several parts along the fault profiles do not show a correlation between depocenters and ST values. For the EBF the depocenter south of Freiburg and near Rastatt do not correlate with higher ST values. Along the WBF, ST is low at the small depocenter south of Colmar and ST is high in the central fault segment, where the graben is very shallow.

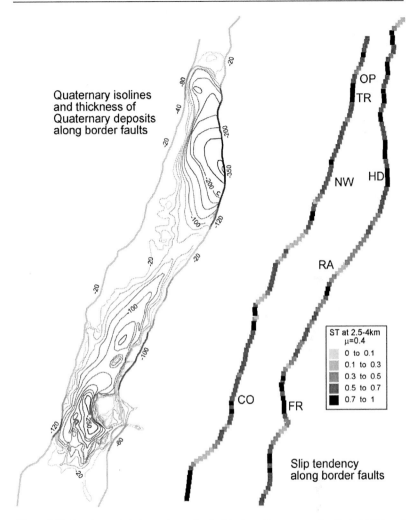

Figure 5.21. Depth map (relative to the surface) of the base Quaternary after Bartz (1974) and Haimberger et al. (2005). Large thickness adjacent to the border faults implies large vertical displacements on them. To the right, slip tendency distribution along the border faults indicates potentially active fault segments. OP = Oppenheim, TR = trench site (Chapter 2), HD = Heidelberg, NW = Neustadt an der Weinstrasse, RA = Rastatt, CO = Colmar, FR = Freiburg im Breisgau.

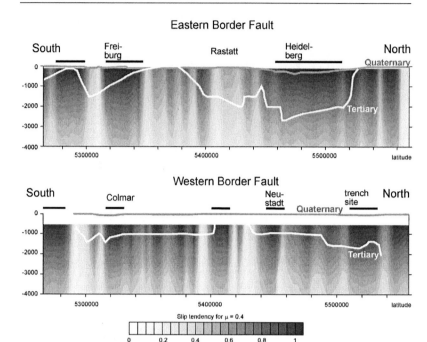

Figure 5.22. Profiles of the Eastern and Western Border Faults. Contours show slip tendency distribution with depth. Fault segments with increased ST are marked with black thick lines. Depths of Tertiary after Doebl and Olbrecht (1974), of Quaternary deposits after Bartz (1974) and Haimberger et al. (2005).

5.8. Discussion

5.8.1. Evaluation of stress field modeling results

In order to predict the fault reactivation potential of URG faults using the ST parameter, emphasis was placed in this study on deriving the best-fit stress field model as input for ST calculations. A good correlation of tectonic regime and stress distribution with depth to reference data from wells, geological and seismological observations was achieved with an elastic 3D FE model (model 2). The model included varying material properties and was loaded in subsequent steps with gravity and lateral displacements. The URG region was modeled as a single crustal block with different materials assigned for the graben fill and the shoulder area. Advantages of this approach chosen for stress field modeling are that in comparison to relatively complex and extensive numerical modeling studies on the URG such as those of Buchmann and Connolly (2007) and Schwarz (2006), the model construction of this work is relatively simple and sensitivity analysis of the loading steps is straightforward.

The results of the stress field model show that the three stress components are in the right order in magnitude for the upper part of the crust. With increasing depth, the maximum horizontal stress S_H becomes lower than the extrapolated reference data and the minimum horizontal stress S_h becomes higher. This results in an underestimation of the differential stress (Figs. 5.11, 5.14). The ST results on the basis of this stress field model show a strong variation with depth with high values near the surface and significantly lower values below 5 km depth (Fig. 5.16A). For typical friction coefficients μ of 0.4 to 0.6 on fault surfaces, the model predicts a stable state and no fault reactivation of optimally oriented faults at larger depths. Considering that the stress field model underestimates the differential stresses, a small increase to more realistic values would still predict a stable state. On the basis of the relatively low seismicity and faulting activity of the URG area, a stable stress state is also expected. Therefore, the stress field modeling and ST results are considered to represent a realistic scenario. Investigation of the fault reactivation potential for the North German Basin and the Lower Rhine Graben predict similar results. For both areas, faults are not likely to be reactivated at hypocenter depths if friction coefficients range between 0.3 to 0.6 (Connolly et al., 2003; Worum et al., 2004; Leydecker, 2005b). The modeling results of Connolly et al. (2003), Worum et al. (2004) and Leydecker (2005b) and the results of this work on URG faults predict slip only for extremely low friction values (μ < 0.3). This demonstrates that none of the modeling studies can explain why faults reactivate and earthquakes nucleate at typical hypocenter depths (8 – 10 km). Thus, the use of too simple friction laws as for the modeling in this thesis seems to be insufficient for predicting the reactivation potential at crustal depths. In order to explain the processes of rupture nucleation more sophisticated models would be required honoring thermal processes, spherical geometries and crustal heterogeneities as well as stress relaxation processes due to faulting activity and creep deformation processes.

5.8.2. *Slip tendency for documented active faults*

The ST for faults with a documented Quaternary faulting history has been analyzed. Most of the active faults exhibit high ST values (Fig. 5.19). This means that these faults are probably being reactivated due to their favorable orientation with respect to the present-day regional stress field. Large vertical displacements, syndepositional movements and surface expressions characterize these active faults. Among the active faults is the WBF scarp segment in the northern URG that has been identified as an active fault segment from paleoseismological trenching and geomorphological analysis (*Chapters 2 to 4*). The high ST results for this segment support its activity. Furthermore, ST results support activity of faults identified on river seismic profiles, of a fault associated with as terrace scarp in the Vorderpfalz and of the *Heidelberger Loch* segment of the EBF (Fig. 5.19). Geomorphological investigations carried out in the framework of this thesis have identified several faults in the Mainz Basin that have a surface expression and affected the landscape owing to their Quaternary reactivation (HTBF, southern Niersteiner Horst fault, Riederbach Valley and Pfrimm Valley Faults). However, ST analysis predicts that these faults are inactive. These faults are oriented NE-SW to E-W, which is not favorable for slip to occur under the present-day stress field. For the HTBF and the Niersteiner Horst a long-term faulting history is documented (Sonne, 1969; Anderle, 1987). With the uplift of the Rhenish Massif and the Mainz Basin, these inherited fault structures of Variscan origin were reactivated and accommodated large amounts of this regional uplift. This shows that fault reactivation is not only related to favorable orientation. With the relative motions of crustal blocks, less favorably oriented faults can also be reactivated. It should also be noted that the simple stress field model used herein is not able to explain such block movements nor does it include the thermal processes that cause the regional uplift of the Rhenish Massif.

5.8.3. Slip tendency results and relation to regional seismicity

Figure 5.1 displays the distribution of seismicity in the URG area that is widely distributed over the graben and the shoulders, in particular in the southern part of the area. The modeling results suggest that the lateral distribution of stresses is related to the topography of the URG area (Fig. 5.13). This relationship accounts also for the differential stress, which is critical for fault reactivation and generation of earthquakes. The stress field model predicts high differential stresses for the elevated shoulder areas (Fig. 5.13). However, these areas do not coincide with earthquake clusters (Fig. 5.1). This highlights that the lateral distribution of seismicity cannot be explained by the distribution of differential stress. Moreover, it becomes evident that the presence of optimally oriented faults and relatively high differential stresses are not the only sources for the seismicity of the URG area.

The reason for the occurrence of earthquakes at typical depths of 8 to 10 km can also not be solved from the ST analysis of this study. The display of the depth distribution of the differential, shear and normal stresses as well as the Mohr circles demonstrate that the stress field model 2 is less critically stressed than the extrapolated data from the Soultz-sous-Forêt site (Figs. 5.11, 5.14, 5.16). This underestimation results in significantly low ST values at typical hypocenter depths and thus the ST analysis predicts that earthquake activity at these depths is rather unlikely. Moreover, it illustrates the complexity of seismic activity and demonstrates the current lack in understanding the occurrence of earthquakes. It also becomes evident that with a ST analysis one cannot predict the location for earthquakes to occur.

However, the method of slip tendency provides a way to assess which fault segments are near the ideal orientation for slip and are the most likely to being reactivated by identification of relative changes of ST. From this ST analysis, it is suggested that for the URG region faults of intermediate ST values (on 0° – 020° and 120° striking faults) coincide with active nodal planes and are therefore seismically active faults. Focal mechanism inversion of two recent seismic events in the northern URG supports the seismic activity of 020° and 120° striking fault surfaces (M. Wagner, pers. comm. 2007). In addition, the trenching study revealed a number of indications for a large seismic event on the 015° striking WBF segment (*Chapter 2*). This large and extremely rare event for this area occurred on a fault segment characterized by intermediate ST values.

Another observation from the ST analysis is that faults with the highest ST values and apparently optimal orientation for fault slip (170°-fault set) are characterized by the largest vertical slip motions documented for geological times scales. However, these faults are not seismically active at present. This leads to suggest that optimally oriented faults in the URG area are presently characterized by aseismic creep. These faults are likely to reactivate and therefore show significant vertical displacements. They are not associated with earthquakes because they might not be able to accumulate sufficient stress. In contrast, faults that are oriented nearly perpendicular to S_H might be capable of loading stresses until they are released during an earthquake. This could be the case for faults with a NE-SW to ENE-WSW strike orientation (065° fault set). The Rastatt 1933 earthquake, which was the largest recorded in the graben (M_L estimated 5.4; Leydecker, 2005a) occurred on a graben parallel fault that is not favorably oriented

(040° strike). The ST analysis predicts the lowest ST for this NE-SW trending fault (green fault segment in Fig. 5.19). In this central graben segment, two other earthquakes with magnitudes 5 occurred on probably NE-SW oriented faults (Forchheim 1948 and Seltz 1952 earthquakes). The HTBF in the northwest of the URG is also a NE-SW striking fault with documented seismicity, although at a relatively low level (maximum intensity I_0 reached was V; Ahorner and Murawski, 1975). These observations demonstrate that larger earthquakes in the area may occur on faults with low ST and orientations perpendicular to S_H. It is interesting to note that faults with the highest ST values are relatively frequent in the URG area (170° and partly 020° fault set), while faults with the lowest values are apparently less frequent (Figs. 5.15, 5.18). This leads to suggest that the large number of optimally oriented faults under present-day stress field conditions (high ST values), which are proposed herein to be mainly creeping faults, reflect the overall low seismicity and the aseismic faulting behavior of the URG area as suggested by Ahorner (1975), Fracassi et al. (2005) and Bonjer et al. (1984).

5.9. Conclusions

This study tries to assess the location of possibly active fault segments in the URG area. In order to investigate this a 3D stress field model has been developed that is used as input for a slip tendency (ST) analysis on URG faults. The ST results are then compared to the observed seismicity and faulting data. It should be noted that the stress field model used in this study is not capable of explaining the spatial distribution of seismicity in the URG area, the processes of earthquake nucleation at depth or predicting the location and timing of future earthquakes.

The 3D stress field model has been loaded with gravity and displacement boundary conditions in order to achieve a realistic 3D stress state. The modeling results reveal that the tectonic regime for the URG and surrounding areas is mainly transtensional for the upper crust. At depths below 12 km and in particular below elevated areas, the regime is extensional. These tectonic regimes are confirmed by fault slip data in outcrops and earthquake data (Plenefisch and Bonjer, 1997; Lopes Cardozo and Behrmann, 2006). The stress field is then used as input for ST analysis. Stress field modeling and ST analysis for URG faults predict that the URG is in a stable state under present-day stress field conditions. A stable state means that damaging earthquakes are rare and deformation rates are low. This is in agreement with the low seismicity of the area (e.g. Ahorner, 1983; Bonjer, 1997) and the low fault slip rates (maximum 0.2 mm/yr for the Quaternary based on data of Bartz, 1974).

Several active faults have been previously identified from seismic interpretations and sediment core correlations (e.g. Bartz, 1974; Haimberger et al., 2005; G. Wirsing, pers. comm., 2005) as well as from trenching and geomorphology in the framework of this thesis (*Chapters 2 to 4*). Geological observation revealed relatively large vertical displacements in particular on 170° striking faults. The ST analysis predicted a high reactivation potential for these faults, since they are favorably oriented under present-day stress field conditions. Nodal planes of this orientation are not documented and larger earthquakes (in the order of magnitude 5 for the URG area) are not associated with these faults. Based on the combined analysis of geological, seismological and modeling data, it is therefore suggested that 170° striking faults are creeping faults. These faults are likely to slip and not associated with earthquakes because they might not be able to accumulate sufficient stress. In contrast, larger earthquakes have been documented for NE-SW oriented faults that strike nearly perpendicular to S_H. Faults with this orientation exhibit the lowest reactivation potential. A plausible explanation for the occurrence of seismicity along such faults is that due to their low ST, stresses can load up on them until they are released during a larger earthquake. Geomorphological and geological analysis document Quaternary activity for several NE-SW oriented faults in the northern URG (*Chapters 3 and 4*). The fault slip might be explained by regional block motions (uplift of Rhenish Massif and Mainz Basin), which cause reactivation of less optimally oriented faults. The ST analysis presented herein demonstrated a way to identify fault segments in the URG area that are prone to reactivation under present-day stress field conditions. By means of a combined analysis of geological, seismological and ST modeling data, a prediction of the faulting behavior is suggested and a separation of URG faults into creeping and seismic faults is proposed. It is clear from the limitations of the modeling approach used that this prediction needs to be investigated further in future studies with the use of more sophisticated models.

CHAPTER 6

SYNTHESIS

The objective of the research presented in this thesis was to assess the active tectonics of the Upper Rhine Graben (URG), with emphasis placed on the northern part of the graben. The period investigated covered the Quaternary and Present. The URG is an area of intensive human modifications and active Quaternary surface processes. On the other hand, neotectonic deformation and the level of seismicity are low. This thesis highlights methods to detect the records of tectonic activity in such an intra-plate setting. The methods listed below have been applied in this thesis and provide information on fault activity and fault characteristics at different spatial and temporal scales:

- paleoseismological trenching,
- shallow geophysical measurements,
- structural fault analysis,
- terrace mapping,
- quantitative geomorphic measurements,
- stress field modeling using 3D finite element (FE) modeling, and
- 3D slip tendency analysis.

This thesis started with fieldwork presented in *Chapter 2*. The aim of the field campaign was to obtain information on the paleoseismicity of the northern URG through trenching. Prior to trenching, geophysical measurements were made at multiple sites on both the eastern and western border of the northern URG (Appendix 1). The integration of results obtained from shallow reflection seismic, geoelectrical resistivity and from ground-penetrating radar measurements indicated near surface deformation along a segment of the WBF. This segment follows a linear morphological scarp of considerable height (50 – 100 m) that suggests a fault scarp morphology. In total 4 trenches were opened across the WBF. These exposed a several meter wide fault zone in Middle Pleistocene to Late Pleistocene fluvial deposits. It could be demonstrated with structural analysis techniques that the fault zone consisted mainly of NNE-SSW

(015°) striking extensional faults. Maximum vertical displacement on individual faults was 0.5 m. Careful investigation permitted to distinguish between tectonic deformation structures and gravitational structures. Although gravitational movements at a small level were observed at for the trench site, these are not considered to have caused the fault zone. Based on the occurrence of liquefaction features, which are typically induced by earthquakes, and the deformation behavior of the trench deposits (brittle deformation of clay-rich sand) it appeared likely that the deformation was associated with tectonic deformation of the WBF. It is concluded that the deformation was seismic and caused by a single earthquake event. Although a number of criteria speak for an earthquake event, aseismic creep movements cannot be fully excluded. Yet, the interpretation that an earthquake is responsible for the observed deformation is considered more likely. Based on the observed displacement, the magnitude of this earthquake is estimated to be M_w 6.5. The deformed deposits were dated and the results point to an earthquake event between 19,000 and 8,000 years ago. Tectonic deformation before or after this period was not detected. Since an earthquake of magnitude 6.5 has not been documented for the northern URG, the findings of the trench site reveal an extremely rare event. Only one single event was identified in the trenches and hence it was not possible to predict any recurrence time for large earthquakes on the WBF. However, it can be assumed that recurrence intervals in this intra-plate setting are in the typical range of tens of thousands of years. The trenching provides detailed information on the WBF activity, however, only at a local scale and for a limited period. In order to gain insight into the long-term behavior of the WBF the geological and geomorphological record of the trench site was investigated. At first, the evolution of tectonic deformation, sediment deposition and erosion at the trench site was reconstructed. The reconstruction demonstrates that tectonics and sedimentary processes interacted with each other. In particular, it is interesting to find that the limit of lateral fluvial erosion coincides with the location of the fault zone. Considering the scarp morphology of the WBF at the trench site, it is concluded that this is the result of the interplay of activity of the WBF and fluvial erosion and sedimentation during the Quaternary.

In order to understand the long-term development of this WBF scarp, the scarp morphology and the landscape of the larger area was investigated (*Chapter 3*). From field surveys on the plateau of the scarp, multiple terrace surfaces were mapped. Results confirmed that the scarp was formed by fluvial incision and aggradation and consisted of a terrace stack. Individual surfaces were grouped and mapped on topographic maps of the scarp and further south along strike of the WBF. It is assumed that the River Rhine system and its tributaries from the Pfälzer Wald formed these terraces. These newly mapped terraces/terrace groups can be correlated with previous terrace mapping on the western border of the northern URG, which enables construction of a longitudinal profile that extends from the Vorderpfalz to the southern parts of the Rhenish Massif. This profile shows that the older terrace groups (higher main and main terraces) were successively uplifted downstream. For the younger terraces (middle and lower terraces), no displacements could be detected. The older terraces are displaced by several individual faults bounding the Mainz Basin. One of the faults, the Riederbach Valley Fault, was previously unknown and its existence is proposed herein for the first time. Using the approximate ages of the terraces, displacement rates were determined, which are in the order of 0.01 – 0.08 mm/yr for the last 800 ka. It is suggested that during the Late Pleistocene, deformation remained at this low level. Displacements since that time cannot be detected because they would be too small to be detected from terrace mapping (in the order of a meter or less). Trenching or coring may be suitable methods to detect surficial deformation. The terrace study shows that potential sites for investigating Late Pleistocene faulting activity are located in the Mainz Bingen Graben and at the Niersteiner Horst.

The record of fluvial terraces was then used to reconstruct the Pliocene to Present evolution of the drainage system of the northern URG (*Chapter 4*). This reconstruction was supplemented with information on sediment distributions and fault displacement data and presented as a series of paleogeomorphic maps (Fig. 4.5). Results show that uplift of the western part of the northern URG can be attributed to the regional uplift of the Rhenish Massif and the Mainz Basin. Uplift of the western margin included reactivation of several graben parallel faults. In contrast, along the Eastern Border Fault (EBF) of the northern URG rapid subsidence resulted in the accumulation of 380 m of Quaternary deposits in the *Heidelberger Loch*. Numerous faults in the northern URG exhibit apparent morphological expressions in the forms of scarps and valleys. It is concluded that activity on these faults has been counteracted by the fluvial dynamics so that fault scarps coincide with terrace scarps. Several composite scarps are newly identified in the northern URG from this work (Fig. 4.11).

Quantitative measurements of geomorphic indices of the configuration of the landscape were used for characterizing lithological or tectonic controls on the morphology and drainage systems and enabled determination of the balance between erosion and tectonic activity (*Chapter 4*). Geomorphic indices calculations indicated increased tectonic activity for the WBF located along the WBF scarp and the northern Pfälzer Wald and for the EBF in the southern part of the Odenwald. On the basis of deformed terraces and exposed fault displacements, Pleistocene deformation rates for the northern URG have been estimated to be in the order of 0.03 mm/yr. During the last 15,000 years, erosional and depositional activity has decreased drastically whilst tectonic activity remained at a constant low level. It is concluded that the tectonic morphology now preserved in the northern URG resulted from long-term, low level tectonic activity and has been preserved because of the decrease of erosional activity during the last 15,000 years.

The last method applied in this thesis was 3D FE modeling that addressed the potential for fault reactivation under present-day stress field conditions (*Chapter 5*). Applying 3D FE techniques, the 3D stress field of the URG and surrounding areas was simulated. The present-day stress field is characterized at the upper crust levels by relatively low horizontal stresses, low differential stresses and a tectonic regime of transtension both in the graben and the shoulder areas. Pore pressure conditions are assumed hydrostatic. This situation has been realistically modeled and used as input for 3D slip tendency (ST) calculations for URG faults. Existing fault data and new data obtained from the trenching and geomorphological studies were incorporated in the ST analysis. Assuming typical fault friction (μ 0.4 – 0.6) under the present-day stress field conditions, the ST results predict that faults in the URG area are in a stable state. Most faults with a long faulting history and documented vertical displacements of tens to hundreds of meters are favorably oriented for reactivation (NNW-SSE) and show the highest ST values. As earthquake data does not support this expected activity, it is concluded that NNW-SSE striking (170°) faults are characterized by aseismic creep. In the central segment of the URG, several larger earthquakes (magnitude 5) occurred on faults striking on average N 065°. As this orientation is not favorable for slip, this could explain the accumulation of stresses and their release by earthquakes rather than by continuous creep. The results of the trenching study in the northern URG, however, show evidence for a large earthquake (magnitude 6) on a NNE-SSW (015°) striking segment of the WBF, which is characterized by intermediate ST values. This demonstrates that in the URG area, fault orientation alone is insufficient to fully explain the occurrence of large earthquakes.

All new data obtained during the preparation of this thesis on the characteristics and locations of active faults are presented in the map given in Figure 6.1. Since this thesis focused on the northern part of the URG, most of the active faults identified are located in this area, whilst for the southern URG the map may not be complete. Under the present-day stress field conditions, all faults in the area have apparently a low reactivation potential. The URG system is in a stable state. With increasing pore pressure and/or reduced fault friction, a large number of faults may become prone to reactivation. This has been demonstrated by the current Deep Heat Mining project of the city of Basel, when fluid injection has triggered the reactivation of a NW-SE trending fracture system with earthquakes up to M_L 3.4 (largest earthquake on 08.12.2006; e.g. see homepage of Swiss seismological survey). Based on the ST analysis of this thesis, it is suggested that most of this reactivation, in particular on NNW-SSE striking faults, will probably involve aseismic creep. Figure 6.1 shows an updated active fault map of the URG and highlights potentially creeping faults and fault segments with documented seismicity.

This thesis demonstrates that a number of methods can be used to obtain information on active tectonics in a slowly deforming area, and that they yield data at different spatial and temporal scales. The integration of results enables quantification of long-term and short-term tectonic processes from local to regional scales. Studies carried out in the context of this thesis started at a local scale and derived data on short-term faulting activity along the WBF from paleoseismological trenching. The scale was extended to long-term tectonic processes on the WBF at a regional scale (entire western border of the northern URG) by applying geomorphological methods. Using modeling techniques, short-term tectonic activity was investigated at a regional scale including the entire URG and surroundings. In the framework of this thesis, it was found that the sequence of the workflow could have also been changed. Starting from regional scale with modeling or geomorphology one could derive information on the location of active faults in a region, which could then be further investigated at a local scale by fieldwork techniques. The advantage of this approach is that multiple potential trench sites could be identified of which the most promising ones can be chosen. The lesson learned from this thesis project is that investigation of active tectonics in a slowly deforming area that has been highly modified by anthropogenic activities requires the application of all techniques and the integration of all data sets available.

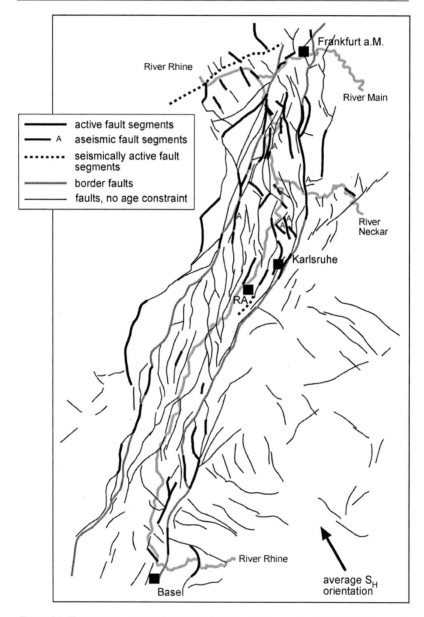

Figure 6.1. The map summarizes all active faults in the URG area identified in this thesis and compiled from other studies (references see caption of Figure 5.4).

APPENDIX

A.1. Earthquakes in the Upper Rhine Graben and Lower Rhine Graben

The earthquakes listed in Table A1.1 are mainly historical earthquakes with estimated intensities I_0 > VII and damages reported. The list has been compiled after Fracassi et al. (2005), Leydecker (2005a) and Schwarz et al. (2006).

No.	Year	Month	Day	Longi-tude	Latitude	Io	Location	Source
1	1021	5	12	7.60	47.50	8.50	Basel	Leydecker 2005
2	1223	1	11	6.83	50.83	7.00	Köln	EKDAG 2006
3	1279	9	2	8.00	49.00	7.00	Elsass	EKDAG 2006
4	1289	9	24	7.80	48.80	7.00	Hagenau (?)	EKDAG 2006
5	1346	11	24	7.60	47.50	8.00	Basel	Leydecker 2005
6	1356	10	18	7.50	47.45	9.00	Basel	Leydecker 2005
7	1357	5	14	7.50	48.17	7.00	Basel	EKDAG 2006
8	1372	6	1	7.60	47.50	7.00	Basel	EKDAG 2006
9	1391	3	23	7.30	47.67	7.00	Sundgau (?)	EKDAG 2006
10	1428	12	13	7.60	47.50	7.00	Basel	EKDAG 2006
11	1504	8	23	6.10	50.80	7.00	Aachen	EKDAG 2006
12	1531	1	26	7.60	47.50	8.00	Basel	Leydecker 2005
13	1531	7	12	6.18	51.37	7.00	Venlo (?)	EKDAG 2006
14	1569	8	6	7.60	47.50	7.00	Basel	EKDAG 2006
15	1610	11	29	7.60	47.50	7.50	Basel	Leydecker 2005
16	1621	5	21	7.30	47.20	7.00	Neuenburg	EKDAG 2006
17	1640	4	4	6.50	50.75	7.50	Dueren	Leydecker 2005
18	1650	9	7	7.60	47.50	7.00	Basel	EKDAG 2006
19	1655	3	29	9.07	48.50	7.50	Tuebingen	Leydecker 2005
20	1669	10	10	7.80	48.60	7.00	Strasbourg	EKDAG 2006
21	1682	5	2	6.52	47.97	8.00	Remiremont	Leydecker 2005
22	1690	12	18	6.20	50.80	7.00	Aachen	EKDAG 2006
23	1721	7	3	7.70	47.45	7.00	Basel	EKDAG 2006
24	1728	8	3	7.92	48.35	7.00	Mahlberg/Lahr	Fracassi et al. 2005, EKDAG 2006
25	1733	5	18	8.02	49.70	7.00	Mainz, Muenchweiler	Leydecker 2005
26	1736	6	12	7.62	47.48	7.00	Basel	EKDAG 2006
27	1737	5	18	8.30	48.90	7.00	Karlsruhe/Rastatt	Leydecker 2005
28	1755	12	18	5.70	50.90	7.00	Maastricht	EKDAG 2006
29	1755	12	26	6.33	50.80	7.50	Dueren	EKDAG 2006
30	1755	12	27	6.25	50.80	7.00	Dueren	EKDAG 2006
31	1756	2	18	6.17	50.78	8.00	Dueren	Leydecker 2005

No.	Year	Month	Day	Longi-tude	Latitude	Io	Location	Source
32	1759	8	23	6.10	50.80	7.00	Aachen (S)	EKDAG 2006
33	1760	1	20	6.40	50.80	7.00	Dueren	EKDAG 2006
34	1776	11	28	7.30	47.77	7.00	Muelhausen	EKDAG 2006
35	1795	12	6	9.42	47.20	7.00	Wildhaus	EKDAG 2006
36	1796	4	20	9.40	47.20	8.00	Rheintal	EKDAG 2006
37	1802	8	23	7.80	48.60	7.00	Strasbourg	EKDAG 2006
38	1846	7	29	7.68	50.15	7.00	St. Goar	EKDAG 2006
39	1869			8.48	49.92	7.00	Gross-Gerau, swarm earthquakes	Leydecker 2005
40	1871	2	10	8.50	49.67	7.00	Lorsch	Leydecker 2005
41	1873	10	22	6.09	50.88	7.00	Herzogenrath	EKDAG 2006
42	1877	6	24	6.08	50.88	8.00	Herzogenrath	Leydecker 2005
43	1878	8	26	6.52	50.93	8.00	Tollhausen	Leydecker 2005
44	1886	10	9	7.92	48.45	7.00	Schutterwald	EKDAG 2006
45	1892	8	9	7.60	50.32	7.00	Koblenz/Boppard	EKDAG 2006
46	1899	2	14	7.65	48.12	7.00	Sasbach	EKDAG 2006
47	1903	3	22	8.17	49.08	7.00	Kandel	Leydecker 2005
48	1910	5	26	7.30	47.40	7.00	Delemont	EKDAG 2006
49	1911	11	16	9.00	48.22	8.00	Ebingen	Leydecker 2005
50	1913	7	20	9.01	48.23	7.00	Albstadt	EKDAG 2006
51	1926	6	28	7.63	48.13	7.00	Sasbach/Kaiserstuhl	Fracassi et al. 2005, EKDAG 2006
52	1933	2	8	8.20	48.85	7.00	Rastatt	Leydecker 2005
53	1935	6	27	9.47	48.03	7.50	Saulgau	Leydecker 2005
54	1943	5	2	8.98	48.27	7.00	Albstadt	EKDAG 2006
55	1943	5	28	8.98	48.27	8.00	Tailfingen	EKDAG 2006
56	1948	6	7	8.33	48.97	7.00	Forchheim	Leydecker 2005
57	1950	3	8	6.72	50.63	7.00	Euskirchen	EKDAG 2006
58	1951	3	14	6.73	50.62	7.50	Euskirchen	Leydecker 2005
59	1952	10	8	7.97	48.90	7.50	Seltz	Leydecker 2005
60	1959	9	4	7.63	48.35	7.00	Plaine De Basse-Alsace (Erstein)	EKDAG 2006
61	1969	2	26	9.01	48.48	7.00	Onstmettingen	EKDAG 2006
62	1970	1	22	9.03	48.30	7.00	Onstmettingen	EKDAG 2006
63	1972	5	18	9.03	48.29	7.00	Onstmettingen	EKDAG 2006
64	1978	9	3	9.03	48.28	7.50	Albstadt	Leydecker 2005
65	1992	4	13	5.92	51.15	7.00	Roermond	Leydecker 2005

Table A1.1. Earthquakes in the Upper Rhine Graben and Lower Rhine Graben with intensities $I_0 > VII$. EKDAG 2006 = Schwarz et al. (2006)

A.2. Geophysical measurements for pre-trenching surveys

Six potential trenching locations in the northern URG have been investigated with geophysical measurements (Fig. A2.1).

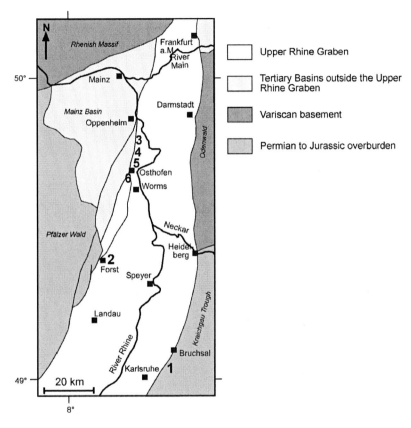

Figure A2.1. Location of potential trenching sites investigated in the northern URG.

A.2.1. Karlsruhe and vicinity

A significant elevation difference of 120 m between the graben shoulders and the River Rhine plain occurs near Karlsruhe. This clear morphological signature of the Eastern Border Fault (EBF) could be an indicator for recent tectonic activity along this fault. Precise levelings, showing subsidence of the graben of 0.9 mm/yr (Zippelt and Mälzer, 1981), and reported normal faulting earthquakes in this region (Karlsruhe-

Forchheim 1948; Ahorner and Schneider, 1974) also point to continuing activity. The focus of geophysical measurements was a valley northeast of Untergrombach, where a linear trending 1.2 m high scarp crossed the valley (Fig. A2.2). The scarp could be the result of a geological boundary between Upper Muschelkalk limestones and loess deposits or the result of human activity. Several ground-penetrating radar (GPR) and geoelectrical tomography profiles did not deliver indications for the presence of a near surface fault along the scarp (Fig. A2.2). The area was not investigated further due to a nearby nature reserve and buildings, which reduced the area available for potential trenching sites.

A.2.2. Forst

At the locality of Forst, a 130 m long profile of up to 8 m height exposes loess deposits and Buntsandstein (Fig. A2.3). This outcrop is one of the few that exposes the boundary between the Pfälzer Wald and the URG. A several meter wide fault zone in intensively weathered sandstones of Buntsandstein steeply bounds the loess deposits. Several extensional faults with vertical displacements of up to 0.5 m occur within the loess. The recent soil is not deformed. Thermoluminescence dating of the loess revealed an age of 369 ka years at the base of the profile and 105 ka years at the top (Weidenfeller and Zöller, 1995). Due to the lack of younger deposits, the timing of deformation could not be closer constrained than the period between 100 ka and present-day. GPR measurements performed on the upper part of the profile (upper terrace) identified the location of the fault zone, but the deeper continuation of the fault could not be detected (Fig. A2.3). This site was not targeted for trenching due to the lack of younger deposits, enabling the dating of the young deformation. Further outcrops with deformed sediments have not been found in the area.

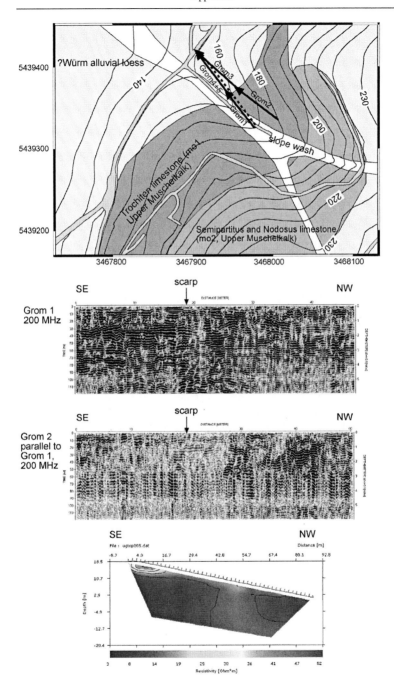

Figure A2.2. Geological map of Untergrombach and location of GPR and geoelectrical profiles. Black lines: GPR profiles. Stippled line: geoelectric profile. Geology after Schnarrenberger (1985). GPR and geoelectrical profiles performed along the EBF near Untergrombach. Coordinates of all maps of the six sites are in Gauss-Krüger coordinate system.

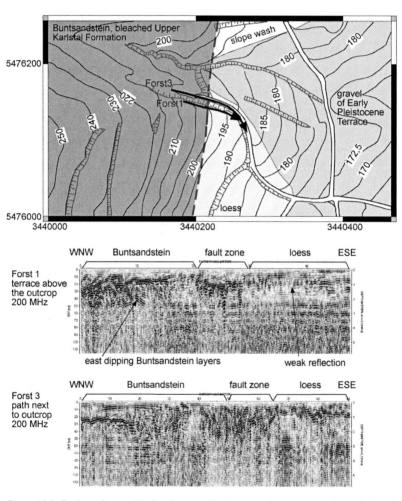

Figure A2.3. Geological map of the locality near Forst and location of GPR profiles. The WBF is indicated with a stippled line. Geology after Manuskriptkarte (1914). GPR profiles from the outcrop near Forst. Topography: Deutsche Grundkarte DK5, Blatt Forst 3440 5476, 1:5,000. Landesamt für Vermessung und Geobasisinformation Rheinland-Pfalz, Koblenz, 1980.

A.2.3. WBF scarp

Geophysical measurements were carried out at four sites along the WBF scarp (Fig. A2.4).

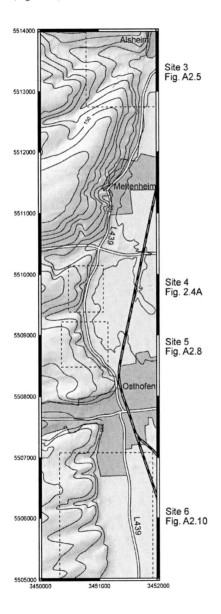

Site 3
Fig. A2.5

Site 4
Fig. 2.4A

Site 5
Fig. A2.8

Site 6
Fig. A2.10

Figure A2.4. Shaded relief map of the WBF scarp with the location of the sites investigated for pre-trenching surveys. Data source: Landesamt für Vermessung und Geobasisinformation, 2002. Coordinates are in Gauss-Krüger coordinate system.

A.2.3.1. Site 3 – Alsheim

Site 3 is located in the central part of the 20 km long scarp near the town of Alsheim. Here, a wide dry valley crosses the scarp. Local deposits cover the valley. At the mouth of the valley a widespread alluvial fan reaches into the plain and covers the lower Würm terrace level (Franke, 2001). Geophysical measurements were performed in the middle of the valley, mainly along paths, since this is where it was expected that the WBF trace crossed the valley (Fig. A2.5). Measurements show that no near surface fault structures occur at site 3 (Fig. A2.6, A2.7). The interpretation of the landform and geophysical data shows that a thick sediment cover, probably fan deposits, masks the upper part of the WBF.

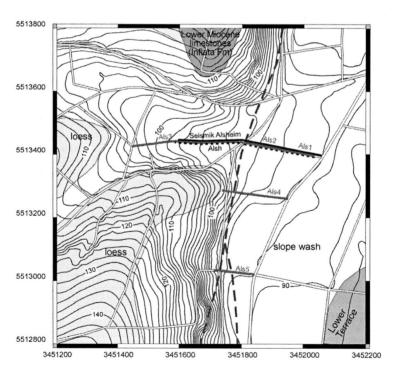

Figure A2.5. Geological map of site 3. Black dashed line: supposed trace of the WBF. Stippled line: geoelectrical profiles. Grey lines: GPR profiles. Black solid line: seismic profile. The

seismic line was performed with the same setup as the seismic line from site 4 (50 kg drop weight, geophone spacing 1 – 2 m; see Chapter 2). Geology and trace of the WBF after Franke (2001).

Next page:

Figure A2.6. GPR profiles from site 3. All profiles are W-E oriented. Shallow cores at profile Als4 revealed topsoil in the upper part and loess with intercalated sand layers in the lower parts of the cores C1 and C2. The layering coincides with dipping reflectors in the GPR profile. No fault pattern is visible in these GPR profiles.

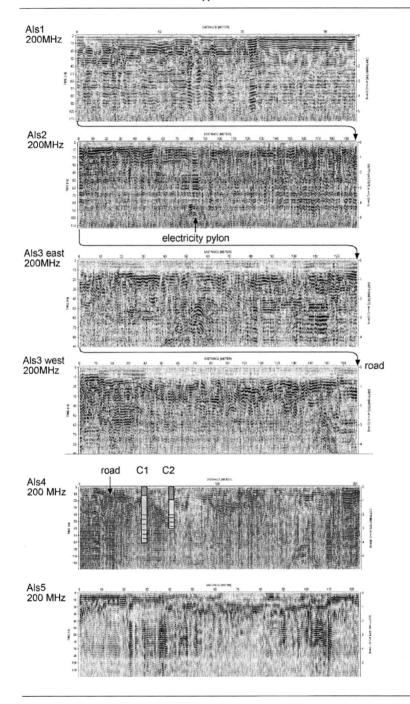

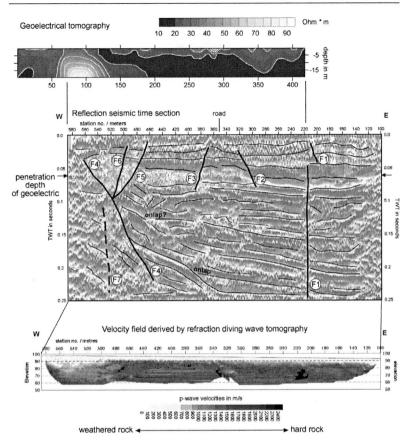

Figure A2.7. Geoelectrical tomography, reflection and refraction seismic lines. Most distinct feature is a low velocity body in the western part of the seismic profiles that coincides with a high resistivity body in the geoelectrical tomography. The feature is bounded by fault 4, which possibly accommodates the largest vertical offset of the faults identified in the seismic line. It remains unclear, which of the faults represent the WBF (fault 1 or 4) or if the WBF lies outside the profile dimension. According to the surface trace mapped in Figure A2.5, the WBF should coincide with the position of the road.

A.2.3.2. Site 4 – Mettenheim

This site was the focus of pre-trenching surveys and based on their measurement results was chosen as trenching site. *Chapter 2* gives a thorough description of the measurements and trenching results.

A.2.3.3. Site 5 – Osthofen, north

Site 5 is located on the southern end of the scarp (Fig. A2.4). At this site, the scarp bends to the southeast and strikes for 1 km length in a NW-SE orientation. The base of the scarp corresponds similar as at the other sites to the lower Würm terrace level (Franke, 2001). Geophysical measurements were performed at the base of the scarp on a fallow vineyard and paths (Fig. A2.8). The measurements revealed no near surface deformation but a structure interpreted as a fluvial channel structure (Fig. A2.9).

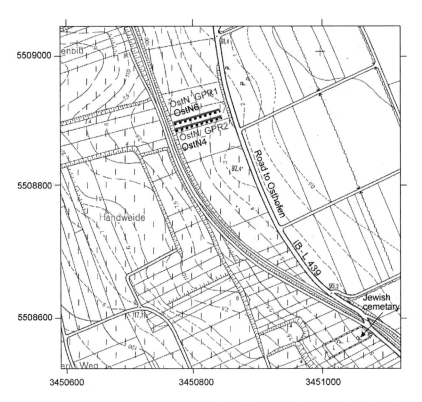

Figure A2.8. Location map of the geophysical profiles at site 5. Topography: Deutsche *Grundkarte DK5, Blatt Osthofen-Nord 3450 5508, 1:5,000. Landesamt für Vermessung und Geobasisinformation Rheinland-Pfalz, Koblenz, 1980.*

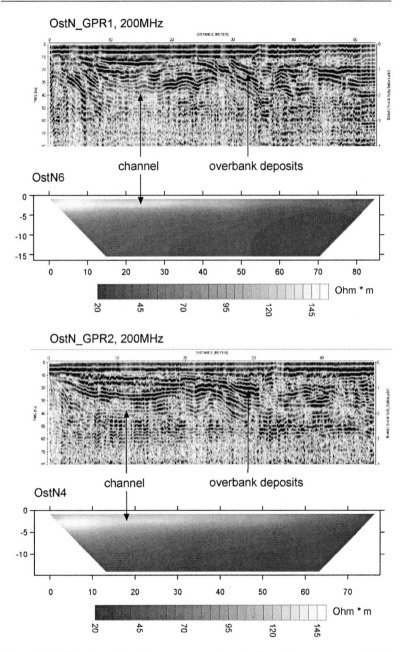

Figure A2.9. GPR and geoelectrical measurements of site 5. All profiles were taken parallel to

each other. Between the northern and the southern profiles is a distance of 6 m. All profiles show a structure interpreted as a river channel.

A.2.3.4. Site 6 – Osthofen, south

Site 6 is situated south of Osthofen along the strike of a linear scarp segment that has an approx. 30 m lower elevation difference than the scarp segment between Osthofen and Oppenheim (50 – 100 m height; Fig. A2.4). GPR and geoelectrical measurements revealed very strong signals and patterns that are interpreted as sedimentary structures, most likely fluvial channel structures (Figures A2.10, A2.11). A near-surface fault has not been detected. Geophysical measurements were performed at the base of the scarp on fallow vineyards and paths (Fig. A2.11). It is suggested that the topography of this southern part of the scarp is the result of fluvial incision.

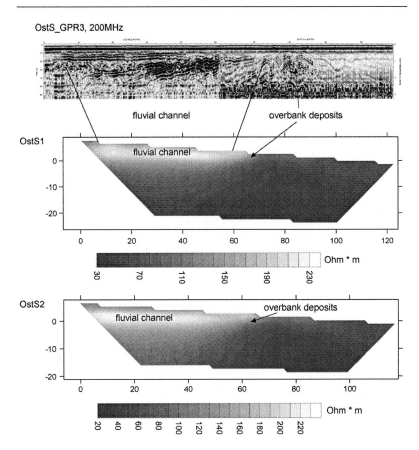

Figure A2.10. GPR and geoelectrical measurements of site 6.

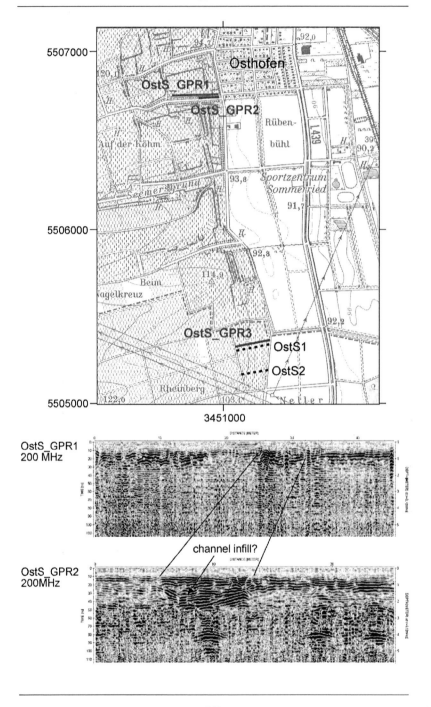

Figure A2.11. Location map of the geophysical profiles at site 6 and GPR profiles performed near the town of Osthofen. Topography: Topographische Karte 1:25,000, Blatt Worms-Pfeddersheim 6315. Landesamt für Vermessung und Geobasisinformation Rheinland-Pfalz, Koblenz, 1996.

A.3.   Stratigraphy of the trenches

Trenching at the site north of Osthofen revealed in total 10 stratigraphic units that are presented in the following.

Unit 1A – sand and clay lenses

Unit 1A consists of sand with lenses of clay-rich sand and clay and is interpreted as a braided river deposit. The unit is exposed at the bottom of TR1 (Fig. 2.6) and has been excavated in a number of cores near TR1 and at TR4 (Fig. 2.9). Core samples adjacent to this trench display strong lateral changes associated with the lenticular bedding. No absolute dating of this unit could be performed, however comparable fluvial quartz-rich sands with interbedded clay have been documented as Pliocene in exposures south of the trenching area (Franke, 2001).

Unit 1B – mottled clay-rich sand

Unit 1B consists of a mixture of clay and sand. Reduction and oxidation patches and abundant manganese nodules throughout the whole unit give it a mottled appearance, typical for post depositional pseudogley formation. The clay-rich sand is interpreted as a still water deposit, possibly an abandoned meander or a lake. Unit 1B is exposed at the bases of TR2 and TR3 (Fig. 2.8). The high kaolin content in the samples of unit 1B indicates Pliocene weathering conditions, so a Pliocene age for unit 1B is most likely (W. Boenigk, E. Hagedorn, pers. comm. 2003).

Unit 2A – cross-bedded sand

Unit 2A consists of cross-bedded, cohesionless sands with gravel layers and is overlying unit 1B with an erosional discontinuity (Fig. 2.8). Unit 2A is interpreted to be a braided river deposit. It is exposed at the western end of TR2 and in the middle and western parts of TR3 (Fig. 2.8). The sands are beyond the dating range of TL, so that no absolute age for unit 2A could be obtained. Correlation of the sands with fluvial terrace levels at the trench site suggests that they form, together with unit 2B, one of the Middle terraces from the Riss glacial period at ~ 450 – 130 ka.

Unit 2B – paleosol

The fluvial sands of unit 2A form the substratum of a soil, unit 2B, consisting of 2 horizons (Fig. 2.8). In the lower C-horizon the clay content increases upward and the original layering is obliterated. Higher clay content and frequent manganese-oxide nodules characterize the 1 m thick B-horizon. Unit 2B is interpreted as a paleocambisol similar, but with a thicker B-horizon, to Pleistocene cambisols seen elsewhere (Scheer and Wourtsakis, 1986; Scheffer and Schachtschabel, 1998). The 1 m thickness of the B-horizon in TR3 is exceptional and implies a long soil formation process. It is likely that this process started in the Eemian interglacial and continued into the Late Holocene.

Unit 3A – conglomerate

Unit 3A is a gravel layer with a matrix of material from unit 1B, unconformably overlying unit 1A in TR1 (Fig. 2.6). Unit 3A is interpreted as a basal conglomerate of the overlying clastic succession. Parts of unit 3A have also been found in cores ~ 10 m west of TR1 on top of the scarp (borehole 52; Figs. 2.4B, 2.11). A TL sample from the base of the overlying unit 3B (MET12) is of > 100 ka.

Unit 3B – interbedded sand and silt

Unit 3B consists of interbedded sand and silt. It is unconformably overlying both unit 3A in TR1 and unit 1B in TR2 (Figs. 2.6, 2.8). The lower 1 m exhibits several deformed zones. The upper 1 m shows parallel lamination. In TR2, unit 3B shows reworking of the underlying unit 1B with embedded clayey sand lenses. Here, diverging strata and sandy deposits dip eastward and indicate high-energy conditions. In this trench, a gradational fining upward transition to unit 4B exists, so that no clear boundary could be determined (boundary in Figure 2.8A is arbitrary). Unit 3B is interpreted as a slope wash deposit. TL dating of a sample from the base of unit 3B (MET12) yields > 100 ka and two samples from the top of the unit (MET13, MET132) yield 19 and 37 ka (Table 2.1, Fig. 2.8).

Unit 4A – iron coated sand and silt

Unit 4A consists of iron coated interbedded sand and silt. The unit is interpreted as the basal part of a second slope wash deposit. It forms an easterly dipping erosional discontinuity with the underlying unit 3B in the western part of TR1 (Fig. 2.6). It has a wedge shape and is located within the fault zone of TR1 (Figures 2.6D, 2.6E). Unit 4A occurs approximately 6 m east of the fault zone at the base of TR1 and in cores exhibiting almost no slope (Fig. 2.6A). TL dating from the western part yields 22 ka (MET81) and 31 ka (MET31; Table 2.1, Fig. 2.6A) ages.

Unit 4B – parallel laminated silt and fine sand

Unit 4B consists of parallel laminated silts and fine sands. The unit is interpreted as a slope wash deposit consisting of fluvially reworked loess and local sands. Due to the high content of loess, it is named 'alluvial loess'. The unit is present in all 4 trenches (Figs. 2.6, 2.8, 2.9). In TR1, unit 4B occurs in the whole 26 m long trench overlying unit 4A (Fig. 2.6). In TR2 and TR3, unit 4B is exposed in the eastern parts of the trenches overlying units 1B and 3B respectively (Fig. 2.8). In TR4 this unit 4B is exposed at the bottom of the trench (Fig. 2.9). TL dates from all 4 trenches have consistent ages for unit 4B ranging from 23 to 14 ka and pointing to a Late Würm age (Table 2.1, Figs. 2.6A, 2.8, 2.9). Due to reliable dating of this unit, a reconstruction of the fault timing has been possible.

Unit 5A – colluvium, unit 5B – recent soil

A recent soil, unit 5B, is present in all 4 trenches consisting of silty and fine sandy material (Figs. 2.6, 2.8, 2.9). In TR4 unit 4A has been replaced by a thick soil (unit 5A), interpreted as a chernozem of possibly Boreal age (8 ka; R. Dambeck and M. Weidenfeller, pers. comm. 2003). In TR2 and TR3 the soil overlies a Late Middle Age colluvium (unit 5A), dated with Late Middle Age pieces of pottery (J. Schütz, pers. comm. 2003), and is therefore younger than this (Fig. 2.8). Conversely, TL dating from the base of the soil in TR1 gives ages from 5 – 10 ka (Fig. 2.6A). It is important to note that the TL ages of unit 5A between 4 and 24 ka (MET115, MET118, MET131; Table 2.1) indicate only the age of the reworked material and not the age of the colluvium.

A.4. Stratigraphy of terraces in the northern URG and Upper Middle Rhine Valley

The following tables summarize the characteristics of the individual terrace units mapped previously in the Vorderpfalz, Pfrimm Valley, Mainz Bingen Graben, Mainz Basin, Lower Main Valley and Upper Middle Rhine Valley.

Terrace level	Height (m a.s.l.)	Relative size	Profile base to top	Composition	Special features	Estimated age
Niederterrasse	90 – 110 (100 – 140)	wide		chert, quartzite, conglomerate, Buntsandstein		Würm
Talwegterrasse, 2 steps at 10 m and 15 m above valley	100 – 117 (110 – 145)	small	reddish sands (Klingbachtal)	Buntsandstein (few bleached), quartzite, chert,	ice-wedges	Riss: correlation of terrace with glacial period, concept of Penck (1910)
Hochterrasse, 2 steps at 20 m and 25 m above valley	115 – 150 (135 – 160)	wide	sands, gravel, clay		ice-wedges, solifluction, cryoturbation	Mindel: correlation of terrace with glacial period
Hauptterrasse	145 – 170 (160 – 250)	interm.	gravel	Buntsandstein (partly bleached, rounded), reworked Pliocene sands, quartzite, chert	cryoturbation	Günz: correlation of terrace with glacial period
Glacis-Niveau	(200 – 280)		course gravel (rounded edges)	Buntsandstein (mostly bleached)		Upper Pliocene

Table A4.1. Stratigraphy of terraces in the Vorderpfalz after Stäblein (1968). Heights are in meter above present-day valley floor after Stäblein (1968). Heights in brackets are surface heights above sea level.

Terrace level	Height (m a.s.l.)	Mean thickness (m)	Relative size	Profile base to top	Composition	Special features	Estimated age
Niederterrasse	2-4	2	small to wide	gravel, alluvial loess, gravel, soil	quartzite, quartz, porphyry, limestone	cryoturbation in alluvial loess	Würm
jüngere Mittelterrasse	6 - 8	3	interm.	gravel, sandy loess, aeolian loess, soil	quartzite, quartz, porphyry		Late Riss
ältere Mittelterrasse	10	4	interm.	gravel, of aeolian loess, soil, alluvial loess	quartzite, quartz, porphyry, Buntsandstein, limestone	solifluction	Early Riss
jüngere Hochterrasse	20	7	wide	gravel and clayey reddish sand, loess or soil cover	quartzite, quartz, porphyry, limestone, Buntsandstein, Schleichsand	ice-wedges	Mindel (II/II – III)
ältere Hochterrasse, untere Stufe	30 - 35	2	small	clayey silt, gravel, loess and soil cover	quartz, quartzite, porphyry, Buntsandstein		Mindel (II)
ältere Hochterrasse, obere Stufe	45	14	small	reworked base, sand and gravel layers with clay-rich sand layers	quartzite, Buntsandstein, iron-rich conglomerate	cryoturbation at base	Mindel (II)
jüngere Hauptterrasse	70 (270 m a.s.l.)	1 – 2	small	gravel covered by thin soil	quartz, porphyry, Buntsandstein, limestone, baryte, phonolite, iron-rich conglomerate	no remnants in Tertiary catchment area: possibly due to landsliding	Mindel (I)
ältere Hauptterrasse	110 (285 m a.s.l.)	2.5	small	gravel, thin loess cover	quartz, porphyry, Buntsandstein, limestone		Günz

Table A4.2. Stratigraphy of terraces along the Pfrimm valley after Leser (1967). Heights are in meter above present-day valley floor after Leser (1967). Heights in brackets are surface heights above sea level.

Terrace level	Height (m a.s.l.)	Mean thickness (m)	Relative size	Profile base to top	Composition	Special features	Estimated age
T1/T2	75 – 89 (68 – 82)	~ 8	wide, max. 2 km	T2 partly covered with loess and aeolian sand, grey, fine-middle sands, carbonate-rich, cross-bedding, poorly sorted	Buntsandstein, quartzite, milk quartz, few Muschelkalk, chert, granite, gneiss	ice-rafted debris of 1 m large Buntsandstein pieces	Würm
T3	85 – 105	12 – 15	wide	loess or aeolian sand cover, reddish, iron-coated gravel, carbonate-rich, intercalated fine, sand lenses	Main provenance: milk quartz, Buntsandstein, chert, few limestone Rhine: granite, gneiss, chert), quartzite, Tertiary limestone	ice-rafted debris of Buntsandstein, Muschelkalk, gravel max. 30 cm diameter	Riss: Riss loess cover on t4 River Main terrace that correlates with T3 Semmel (1966), Kandler (1970)
T4	117 – 126		small on southern side, wide on northern side	loess or aeolian sand cover, iron-coated, rounded gravel (max. 20 cm), carbonate free, frequent Buntsandstein plates	Buntsandstein, quartzite, chert, few Muschelkalk, Tertiary limestone, gneiss, granite, very few basalt, carbonate free	ice-rafted debris of Buntsandstein (max. 1 m x 0.2 m), solifluction debris	Mindel: correlation with T3 River Main terrace Semmel (1969)
T5	135 – 140		small	gravel	quartzite, phyllite, Buntsandstein, chert, limestone		
T6 (Mosbach sands)	115 – 145	13-15 m Middle and Upper Mosbach, 1-5 m Lower Mosbach	wide near Mainz/Wiesbaden, small along the graben	Middle and Upper Mosbach: grey, carbonate-rich, fine-middle sands, intercalated gravel lenses Lower Mosbach: iron-coated, carbonate free gravel	Middle and Upper Mosbach: Buntsandstein, quartzite, Hermeskeilsandstone, phyllites, gneiss, granite, Spessart sandstone Lower Mosbach: Taunus gravel, Main gravel, Alpine Numulite limestone, Muschelkalk, Buntsandstein	Middle/Upper Mosbach: Pleistocene mammalia of warm and cold climate, no morphological terrace, ice-rafted debris of Muschelkalk, Buntsandstein, cryoturbation, ice-wedges	Cromer: pollen analysis & paleo-magnetics at Werlau Middle Rhine Valley and Dyckerhoff quarry in Wiesbaden/ Mainz Bingen graben Bibus and Semmel (1977)
T7	165 – 200		small on southern side, wide on northern side	sands and gravel	Taunus material, Buntsandstein, chert, reworked Pliocene (carbonate-rich sands),	heterogeneous composition on both sides of Mainz Bingen graben, ice-rafted debris of Buntsandstein	Early Pleistocene: based on topographic position
Early Pleistocene terrace	220 – 235		wide on southern plateau, small on northern side	grey, carbonate-rich sands, intercalated gravel of Buntsandstein		no clear distinction, Early Pleistocene terrace = gravel between T7 and Pliocene	

Table A4.3. Stratigraphy of terraces along the Mainz Bingen Graben after Kandler (1970).

Terrace level *	Height (m a.s.l.)	Mean thickness (m)	Relative size	Profile base to top	Composition	Special features	Estimated age
Lower Terrace	min. 0.5 – 2.5		iterm.		gravel of Tertiary sediments from Mainz Basin, melaphyre, cyrena marls, Schleichsand, marine sands	*Elephas primigenius,* Rhinoceros antiquitatis, Unio ictorum	Würm
Mittelterrasse (Middle Terraces)	120 – 147		small	gravels	Limestone of corbicula unit, Schleichsand, cyrena marls, Pliocene gravel, Bohnerze, melaphyre		
Jüngere Hauptterrasse (Main Terraces)	160 – 180		small	gravels	melaphyr, milk quartz, limetone of corbicula unit, Schleichsand, cyrena marls, Pliocene sands		
ältere Hauptterrasse (Higher Main Terraces)	180 – 260	~ 0.4	small	gravels		tooth of *Elephas meridionalis*	Early Pleistocene age due to tooth

*Table A4.4. Stratigraphy of terraces along the River Selz after Wagner (1931b). *Classification in brackets after proposition of this study.*

Terrace level	Height (m a.s.l.)	Mean thickness (m)	Relative size	Profile base to top	Composition	Special features	Estimated age
t7	85 – 90	max. 5	small	gravel, sandy towards the top	frequent greenish sandstones,	cryoturbation	Würm
t6	90 – 95	max. 5	small	reddish brown gravel, loess cover, aeolian sand, loam	Muschelkalk, Buntsandstein, chert, basalt	teeth of *Elephas primigenius*	Würm?
t5	95 – 100	max. 5	small	gravel covered with loess			Late Riss
t4	105 – 110	max. 5	small	light reddish gravel, covered with loess		Fe-Mn-crusts, few ice-rafted debris of Buntsandstein	Early Riss
t3	107 – 125	max. 10	small	reddish gravel, intercalated sand layer, covered with loess	Buntsandstein, chert, basalt	frequent ice-rafted debris of Bunt-sandstein	Mindel?
t2	120 – 150	5	wide	brownish sandy gravel			Mindel or Günz
t1	60 – 150	~ 25	wide	course, grey, greenish middle grain sands, partly carbonate-rich, intercalated gravel layers	Buntsandstein, chert, basalt, "typical" Pleistocene assemblage	ice-rafted debris of sandstone	Tegelen and Waal
Early Pleistocene sands				brownish, gravelly sands	high epidote content		
Pliocene sands and gravel							

Table A4.5. Stratigraphy of terraces in the Lower Main Valley after Semmel (1969).

Terrace level	Height (m a.s.l.)	Mean thickness (m)	Relative size	Profile base to top	Composition	Special features	Estima-ted age
tR$_{11}$ (younger lower terrace)	69	1 – 4		gravel, sand, debris, few volcanic ash material			
tR$_{10}$ (older lower terrace)	73	2 – 6					
tR$_9$ (lower middle terrace)	101	3 – 7	small				
tR$_8$ (intermediate Middle terrace)	120	4 – 6	small	gravel			
tR$_7$ (upper middle terrace)	139	1 – 3	small				
tR$_6$ (uppermost middle terrace)	203	1 – 3	small				
tR$_5$ (younger main terrace)	190 - 200		wide	gravel, max. 18 m thick loess cover with minimum 3 paleosols	carbonate-free		
tR$_4$ (Mosbach facies)	215 - 220	~ 20 m	small	carbonate-rich gravel with intercalated fine sand layers, increase of sand layers towards the top, partly intercalated paleosol, half-bog soil, loess with 50 cm basaltic tuff, Holocene soil cover	carbonate-rich	carbonate crusts	
tR$_3$ (older main terrace)	235 - 245	max. 8	medium	gravel, loess cover	schist, Horn-blende, garnet	ice-rafted debris, yellow weathering zone	
tR$_2$ (lower Early Pleistocene terrace)	255 - 265	4	medium	gravel	schist, less stable minerals, epidot-zoisite	red weathering zone near surface	
tR$_1$ (upper Early Pleistocene terrace)	275 - 285	few meters	small	sand	quartz, quartzite		
tjP (lower Pliocene terrace)	295 - 305	few meters	wide	gravel	stable heavy minerals		

Table A4.6. Stratigraphy of terraces in the Upper Middle Rhine Valley after Bibus and Semmel (1977).

REFERENCES

Abele, G., 1977. Morphologie und Entwicklung des Rheinsystems aus der Sicht des Mainzer Raumes. Mainzer geogr.Stud., 11: 245-258.

Ahorner, L., 1975. Present-day stress field and seismotectonic block movements along major fault zones in Central Europe. Tectonophysics, 29(1-4): 233-249.

Ahorner, L., 1983. Historical Seismicity and Present-Day Microearthquake Activity of the Rhenish Massif, Central Europe. In: J.H. Illies et al. (Editors), Plateau Uplift: The Rhenish Massif - A Case History. Springer Verlag, Heidelberg, pp. 198-221.

Ahorner, L., 1992. Seismologisches Gutachten zur Erdbebengefährdung des Standortes für die geplante Reststoffdeponie Nord bei Worms-Abenheim. Archiv GLA 6215/32 (Gundersheim), Geologisches Landesamt Rheinland-Pfalz, Mainz.

Ahorner, L., 2001. Abschätzung der statistischen Wiederkehrperiode von starken Erdbeben im Gebiet von Köln auf Grund von geologisch-tektonischen Beobachtungen an aktiven Störungen, DGG Mitteilungen, pp. 2-9.

Ahorner, L., Baier, B. and Bonjer, K.-P., 1983. General pattern of seismotectonic dislocation and the earthquake-generating field in central Europe between the Alps and the North Sea. In: J.H. Illies et al. (Editors), Plateau Uplift: The Rhenish Massif - A Case History. Springer Verlag, Heidelberg, pp. 187-197.

Ahorner, L. and Murawski, H., 1975. Erbebentätigkeit und geologischer Werdegang der Hunsrück-Südrand-Störung. Z. dt. geol. Ges., 126: 63-82.

Ahorner, L. and Rosenhauer, W., 1978. Seismic risk evaluation for the Upper Rhine graben and its vicinity. J. Geophys., 44: 481-497.

Ahorner, L. and Schneider, G., 1974. Herdmechanismen von Erdbeben im Oberrheingraben und in seinen Randgebirgen. In: K. Fuchs and J.H. Illies (Editors), Approaches to Taphrogenesis, Stuttgart, pp. 104-117.

Anderle, H.J., 1974. Block tectonic interrelations between Northern Upper Rhine Graben and Southern Taunus Mountains. In: K. Fuchs and J.H. Illies (Editors), Approaches to Taphrogenesis, Stuttgart, pp. 243-253.

Anderle, H.J., 1987. The evolution of the South Hunsrück and Taunus Borderzone. Tectonophysics, 137: 101-114.

Anderson, E.M., 1905. The dynamics of Faulting. Edinburgh Geological Society Transactions, 8(3): 387-402.

Anderson, E.M., 1951. The dynamics of faulting and dyke formation with applications to Britain. Oliver & Boyd, Edinburgh, 206 pp.

Andres, J., 1958. Geologische und geophysikalische Untersuchungen im Saar-Nahe-Trog. Erdöl und Kohle, 11: 441-450.

Andres, J. and Schad, A., 1959. Seismische Kartierung von Bruchzonen im mittleren und nördlichen Teil des Oberrheintalgrabens und deren Bedeutung für die Ölansammlung. Erdöl und Kohle, 12: 323-334.

Andres, W. and Preuss, J., 1983. Erläuterungen zur Geomorphologischen Karte 1:25000 der Bundesrepublik Deutschland, GMK 25, Blatt 11, 6013 Bingen. Geo Center, Berlin.

Azor, A., Keller, E.A. and Yeats, R.S., 2002. Geomorphic indicators of active fold growth: South Mountain-Oak Ridge anticline, Ventura basin, southern California. Geological Society of America Bulletin, 114(6): 745-753.

Bada, G., Horváth, F., Cloetingh, S., Coblentz, D.D. and Tóth, T., 2001. Role of topography-induced gravitational stresses in basin inversion: The case study of the Pannonian basin. Tectonics, 20(3): 343-363.

Baier, B. and Wernig, J., 1983. Microearthquake Activity near the Southern Border of the Rhenish Massif. In: J.H. Illies et al. (Editors), Plateau Uplift: The Rhenish Massif - A Case History. Springer-Verlag, Heidelberg, pp. 222-227.

Barsch, D. and Mäusbacher, R., 1979. Erläuterungen zur Geomorphologischen Karte 1:25000 der Bundesrepublik Deutschland, GMK 25, Blatt 3, 6417 Mannheim-Nordost. Geo Center, Berlin.

Bartz, J., 1936. Das Unterpliocän in Rheinhessen. Jber. Mitt. oberrhein. geol. Ver., N.F., 25: 121-228.

Bartz, J., 1950. Das Jungpliozän im nördlichen Rheinhessen. Notizbl. hess. Landesamt Bodenforsch., VI(1): 201-243.

Bartz, J., 1961. Die Entwicklung des Flußnetzes in Südwestdeutschland. Jh. geol. Landesamt Baden-Württemberg, 4: 127-135.

Bartz, J., 1974. Die Mächtigkeit des Quartärs im Oberrheingraben. In: K. Fuchs and J.H. Illies (Editors), Approaches to Taphrogenesis. E. Schweitzerbart'sche Verlagsbuchhandlung, Stuttgart, pp. 78-87.

Beck, N., 1989. Periglacial glacis (pediment) generations at the western margin of the Rhine Hessian Plateau. In: F. Ahnert (Editor), Landforms and landform evolution in West Germany, pp. 189-197.

Becker, A. and Davenport, C.A., 2003. Rockfalls triggered by the AD 1356 Basle Earthquake. Terra Nova, 15(4): 258-264.

Becker, A., Davenport, C.A. and Giardini, D., 2002. Palaeoseismicity studies on end-Pleistocene and Holocene lake deposits around Basle, Switzerland. Geophys. J. Int., 149: 659-678.

Behnke, C., Cloos, H. and Dürbaum, H., 1967. Remarks Concerning the Tectonics of the Upper Rhinegraben. Abh. geol. L.-Amt Baden-Württ., 6, The Rhinegraben Progress Report 1967: 3-4.

Bibus, E. and Semmel, A., 1977. Über die Auswirkung quartärer Tektonik auf die altpleistozänen Mittelrhein-Terrassen. Catena, 4: 385-408.

Birkenhauer, J., 1973. Zur Chronologie, Genese und Tektonik der plio-pleistozänen Terrassen am Mittelrhein und seinen Nebenflüssen. Z. Geomorph. N. F., 17(4): 489-496.

Boenig, W., 1978. Gliederung der altquartären Ablagerungen in der Niederrheinischen Bucht. Fortschr. geol. Rheinld. u. Westph., 28: 135-212.

Boenig, W., 1987. Petrographische Untersuchungen jungtertiärer und quartärer Sedimente am linken Oberrhein. Jber. Mitt. oberrhein. geol. Ver., N.F., 69: 357-394.

Boenigk, W. and Frechen, M., 2006. The Pliocene and Quaternary fluvial archives of the Rhine system. Quaternary Science Reviews, 25(5-6): 550-574.

Boenigk, W. and Hoselmann, C., 1991. Zur Genese der Hönninger Sande (unterer Mittelrhein). Eiszeitalter u. Gegenwart, 41: 1-15.

Bogaard, P.J.F. and Wörner, G., 2003. Petrogenesis of Basanitic to Tholeiitic Volcanic Rocks from the Miocene Vogelsberg, Central Germany. Journal of Petrology, 44(3): 569-602.

Boigk, H. and Schöneich, H., 1970. Die Tiefenlage der Permbasis im nördlichen Teil des Oberrheingrabens. In: J.H. Illies and S. Mueller (Editors), Graben Problems, Stuttgart, pp. 45-55.

Bonjer, K.-P., 1997. Seismicity pattern and style of seismic faulting at the eastern borderfault of the southern Rhine Graben. Tectonophysics, 275: 41-69.

Bonjer, K.-P., Gelbke, C., Gilg, B., Rouland, D., Mayer-Rosa, D. and Massinon, B., 1984. Seismicity and dynamics of the Upper Rhinegraben. J. Geophysics, 55: 1-12.

Bosum, W. and Ullrich, H.J., 1970. Die Flurmagnetometermessung des Oberrheingrabens und ihre Interpretation. Geologische Rundschau, 59: 83-106.

Bott, M.H.P., 1959. The mechanisms of oblique slip faulting. Geological Magazine, 96: 109-117.

Breyer, F. and Dohr, G., 1967. Bemerkungen zur Stratigraphie und Tektonik des Rheintal-Grabens zwischen Karlsruhe und Offenburg. Abh. geol. L.-Amt Baden-Württ., 6, The Rhinegraben Progress Report 1967: 42-43.

Brudy, M. and Zoback, M.D., 1999. Drilling-induced tensile wall-fractures: implications for determination of in-situ stress orientation and magnitude. International Journal of Rock Mechanics & Mining Sciences, 36(2): 191-215.

Brudy, M., Zoback, M.D., Fuchs, K., Rummel, F. and Baumgärtner, J., 1997. Estimation of the complete stress tensor to 8 km depth in the KTB scientific drill holes: Implications for crustal strength. Journal Geophysical Research, 102: 18453-18475.

Brun, J., Wenzel, F. and ECORS-DEKORP team, 1991. Crustal-scale structure of the southern Rhinegraben from ECORS-DEKORP seismic reflection data. Geology, 19: 758-762.

Brun, J.P., 1999. Narrow rifts versus wide rifts: inferences for the mechanics of rifting from laboratory experiments. Philosophical Transactions of the Royal Society A: Mathematical, Physical and Engineering Sciences, 357(1753): 695-712.

Brüning, H., 1975. Paläogeographisch-ökologische und quartärmorphologische Aspekte im nördlichen und nordöstlichen Mainzer Becken. Mz. Naturw. Arch., 14: 5-91.

Brüning, H., 1977. Zur Oberflächengenese im zentralen Mainzer Becken. Mainzer geogr.Stud., 11: 227-243.

Brüning, H., 1978. Zur Untergliederung der Mosbacher-Terrassenabfolge und zum klimatischen Stellenwert der Mosbacher Tierwelt im Rahmen des Cromer-Komplexes. Mz. Naturw. Arch., 16: 143-190.

Brüstle, A., 2002. Morphostructural analyses of the region of Freiburg i. Br. (Upper Rhine Graben). Diploma Thesis, 83 pp.

Buchmann, T.J. and Connolly, P.T., 2007. Contemporary kinematics of the Upper Rhine Graben: a 3D finite element approach. Global and Planetary Change, 58: 287-309.

Büdel, J., 1977. Klima-Geomorphologie. Gebrüder Bornträger, Berlin, 304 pp.

Bull, B.W., 1991. Geomorphic responses to climate change. Oxford University Press, 323 pp.

Bull, B.W. and McFadden, L.D., 1977. Tectonic geomorphology north and south of the Garlock Fault, California. In: D.O. Doehring (Editor), Geomorphology in arid regions, Binghamton, N.Y., State University of New York at Binghamton, pp. 115-138.

Burbank, D.W. and Anderson, R.S., 2001. Tectonic Geomorphology. Blackwell Science, 274 pp.

Byerlee, J., 1978. Friction of Rocks. Pure and Applied Geophysics, 116(4-5): 615-626.

Camelbeeck, T., Van Eck, T., Pelzing, R., Ahorner, L., Loohuis, J., Haak, H.W., Hoang-Trong, P. and Hollnack, D., 1994. The 1992 Roermond earhtquake, the Netherlands, and its aftershocks. Geologie en Mijnbouw, 73: 181-197.

Chen, Y.-C., Sung, Q. and Cheng, K.-Y., 2003. Along-strike variations of morphotectonic features in the Western Foothills of Taiwan: tectonic implications based on stream-gradient and hypsometric analysis. Geomorphology, 56(1-2): 109-137.

Cloetingh, S., Burov, E. and Poliakov, A., 1999. Lithospheric folding: Primary response to compression? (from central Asia to Paris Basin). Tectonics, 18: 1064-1083.

Cloetingh, S. and Burov, E.B., 1996. Thermomechanical structure of European continental lithosphere: constraints from rheological profiles and EET estimates. Geophysical Journal International, 124: 695-723.

Cloetingh, S., Cornu, T., Ziegler, P.A. and Beekman, F., 2006. Neotectonics and intraplate continental topography of the northern Alpine Foreland. Earth-Science Reviews, 74(3-4): 127.

Cloetingh, S., Ziegler, P., Cornu, T. and GROUP, E.W., 2003. Investigating environmental tectonics in northern Alpine foreland of Europe. EOS, Transactions, American Geophysical Union, 84(36): 349, 356-357.

Cloetingh, S., Ziegler, P.A., Beekman, F., Andriessen, P.A.M., Matenco, L., Bada, G., Garcia-Castellanos, D., Hardebol, N., Dèzes, P. and Sokoutis, D., 2005. Lithospheric memory, state of stress and rheology: neotectonic controls on Europe's intraplate continental topography. Quaternary Science Reviews, 24: 241-304.

Cloetingh, S.A.P.L. and Cornu, T.G.M. (Editors), 2005. Neotectonics and Quaternary fault-reactivation in Europe's intraplate lithosphere. Quaternary Science Reviews, 24 (3-4), 235-508.

Cloos, H., 1937. Ergebnisse regionaler Schweremessungen im Oberrheintal mit Bemerkungen zur gravimetrischen Struktur Süddeutschlands. Oel und Kohle, 13: 1065-1073.

Connolly, P., Gölke, M., Bässler, H., Fleckenstein, P., Hettel, S., Lindenfeld, M., Schindler, A., Theune, U. and Wenzel, F., 2003. Finite Elemente Modellrechnungen zur Erklärung der Auffächerung der größeren horizontalen Hauptspannungsrichtung in Norddeutschland - Endbericht -, Geophysical Institute, Karlsruhe.

Cornell, A., 1968. Engineering seismic risk analysis. Bull. Seism. Soc. Am., 58: 1583-1606.

Cornet, F.H., Bérard, T. and Bourouis, S., 2007. How close to failure is a granite rock mass at a 5 km depth? International Journal of Rock Mechanics & Mining Sciences, 44: 47-66.

Cornu, T.G.M. and Bertrand, G., 2005. Numerical backward and foreward modelling of the southern Upper Rhine Graben (France-Germany border): new insights on tectonic evolution of intracontinental rifts. Quaternary Science Reviews, 24(3-4): 353-361.

Cuong, N.Q. and Zuchiewicz, W.A., 2001. Morphotectonic properties of the Lo River Fault near Tam Dao in North Vietnam. Natural Hazards and Earth System Sciences, 1: 15-22.

Cushing, M., Lemeille, F., Cotton, F., Grellet, B., Audru, J.-C. and Renardy, F., 2000. Paleo-earthquakes investigations in the Upper Rhine Graben in the framework of the PALEOSIS project.

Dambeck, R. and Bos, J.A.A., 2002. Late glacial and Early Holocene landscape evolution of the northern upper Rhine river valley, south-western Germany. Z. Geomorph. N.F. Suppl., 128: 101–127.

Dambeck, R. and Thiemeyer, H., 2002. Fluvial history of the northern Upper Rhine River (southwestern Germany) during the Lateglacial and Holocene times. Quat. Intern., 93-94: 53-63.

Davy, P. and Cobbold, P.R., 1991. Experiments on shortening of a 4-layer model of the continental lithosphere. Tectonophysics, 188: 1-25.

Deichmann, N., 1990. Seismizität der Nordschweiz 1987-1989, und Auswertung der Erdbebenserien von Grünsberg, Läufelfingen und Zeglingen. 90-46, Schweizerischer Erdbebendienst Nagra, Zürich.

Delouis, B., Haessler, H., Cisternas, A. and Rivera, L., 1993. Stress tensor determination in France and neighbouring regions. Tectonophysics, 221(3-4): 413-437.

Delvaux, D., Moeys, R., Stapel, G., Petit, C., Levi, K., Miroshnichenko, A., Ruzhich, V. and San'kov, V., 1997. Paleostress reconstructions and geodynamics of the Baikal region, Central Asia, Part 2. Cenozoic rifting. Tectonophysics, 282: 1-38.

DeMets, C., Gordon, R.G., Argus, D.F. and S., S., 1990. Current Plate Motions. Geophysical Journal International, 28: 2121-2124.

Demoulin, A., Pissart, A. and Zippelt, K., 1995. Neotectonic activity in and around the southwestern Rhenish shield (West Germany): indications of a levelling comparison. Tectonophysics, 249: 203-216.

Derer, C.E., 2003. Tectono-sedimentary evolution of the northern Upper Rhine Graben (Germany), with special regard to the early syn-rift stage. PhD thesis, University of Bonn, Bonn, 99 pp.

Derer, C.E., Kosinowski, M., Luterbacher, H.P., Schäfer, A. and Süß, M.P., 2003. Sedimentary response to tectonics in extensional basins: the Pechelbronn Beds (Late Eocene to early Oligocene) in the northern Upper Rhine Graben, Germany. Geol. Soc. Spec. Publ. London, 208: 55-69.

Dezayes, C., Genter, A. and Gentier, S., 2004. Fracture Network of the EGS Geothermal Reservoir at Soultz-sous-Forêts (Rhine Graben, France), Geothermal Energy - The Reliable Renewable. GRS Transactions, Indian Wells, California - USA, pp. 213-218.

Dèzes, P., Schmid, S.M. and Ziegler, P.A., 2004. Evolution of the European Cenozoic Rift System: interaction of the Alpine and Pyrenean orogens with their foreland lithosphere. Tectonophysics, 389: 1-33.

Dèzes, P. and Ziegler, P.A., 2002. Moho depth map of Western and Central Europe, World

Wide Web Address: http://www.unibas.ch/eucor-urgent.

DGM5, 2002. Digitales Geländemodel DGM5. Landesamt für Vermessung und Geobasisinformation Rheinland-Pfalz, Koblenz.

DGM50, 2001. Digitales Geländemodell 1:50000 M745. Vermessungsverwaltungen der Länder und BKG 2001.

Dirkzwager, J.B., 2002. Tectonic modelling of vertical motion and its near surface expression in The Netherlands. PhD thesis, Vrije Universiteit, Amsterdam, 156 pp.

DK5, B., 1980. Deutsche Grundkarte 1:5000 DK5, Blatt Osthofen-Nord, 3450 5508. Landesamt für Vermessung und Geobasisinformation Rheinland-Pfalz, Koblenz.

Doebl, F., 1967. The Tertiary and Pleistocene sediments of the northern and central part of the Upper Rhinegraben. Abh. geol. L.-Amt Baden-Württ., 6: 48-54.

Doebl, F. and Olbrecht, W., 1974. An Isobath of the Tertiary Base in the Rhinegraben. In: K. Fuchs and J.H. Illies (Editors), Approaches to Taphogenesis, Stuttgart, pp. 71-72.

Dziewonski, A.M., Chou, T.A. and Woodhouse, J.H., 1981. Determination of earthquake source parameters from waveform data for studies of global and regional seismicity. Journal Geophysical Research, 86: 2825-2852.

Eitel, B., 2003. Geomorphogenese zwischen Rhein und Neckar: Stand der Forschung und offene Fragen. In: W. Schirmer (Editor), GeoArchaeoRhein: Landschaftsgeschichte im Europäischen Rheinland. GeoArchaeoRhein. Lit Verlag, Münster, pp. 127-152.

Ellis, S., Beavan, J., Eberhart-Phillips, D. and Stockhert, B., 2006. Simplified models of the Alpine Fault seismic cycle: stress transfer in the mid-crust. Geophysical Journal International, 166(1): 386-402.

Ellwanger, D., Lämmermann-Bartherl, J. and Neeb, I., 1995. Baden-Württemberg. In: L. Benda (Editor), Das Quartär Deutschlands. Gebrüder Borntraeger, Berlin - Stuttgart, pp. 255-295.

ENTEC, homepage: http://www.geo.vu.nl/users/entec.

EUCOR-URGENT, Homepage: http://comp1.geol.unibas.ch/.

Ferrill, D.A., Winterle, J., Wittmeyer, G., Sims, D., Colton, S., Armstrong, A. and Morris, A.P., 1999. Stressed Rock Strains Groundwater at Yucca Mountain, Nevada. GSA Today, 9(5): 1-8.

Ferry, M., Meghraoui, M., Delouis, B. and Giardini, D., 2005. Evidence for Holocene palaeoseismicity along the Basel-Reinach active normal fault (Switzerland): a seismic source for the 1356 earthquake in the Upper Rhine graben. Geophys. J. Int., 160(2): 554-572.

Fetzer, K.D., Larres, K., Sabel, K.-J., Spies, E. and Weidenfeller, M., 1995. Hessen, Rheinland-Pfalz, Saarland. In: L. Benda (Editor), Das Quartär Deutschlands. Gebrüder Borntraeger, Berlin-Stuttgart, pp. 220-254.

Field, E.H., Jackson, D.D. and Dolan, J.F., 1999. A mutually consistent seismic-hazard source model for Southern California. Bull. Seism. Soc. Am., 89: 559-578.

Fracassi, U., Nivière, B. and Winter, T., 2005. First appraisal to define prospective seismogenic sources from historical earthquake damages in southern Upper Rhine Graben. Quaternary Science Reviews, 24: 401-423.

Fraefel, M., Densmore, A.L., Ustaszewski, K. and Schmid, S.M., 2006. Geomorphological evidence for recent tectonic deformation in northern Switzerland, Geophysical Research Abstracts, European Geosciences Union 2006.

Franke, W.R., 1989. Tectonostratigraphic units in the Variscan Belt of Europe. Geological Society of America Bulletin, 230: 67-90.

Franke, W.R., 1999. Geologische Karte von Rheinland-Pfalz 1:25000, Erläuterungen, Blatt 6214 Alzey. Geologisches Landesamt Rheinland-Pfalz, Mainz.

Franke, W.R., 2001. Geologische Karte von Rheinland-Pfalz 1:25000, Erläuterungen, Blatt 6215 Gau-Odernheim. Geologisches Landesamt Rheinland-Pfalz, Mainz.

Franke, W.R. and Anderle, H.J., 2001. Geologische Übersichtskarte 1:200000, CC 6310 Frankfurt a.M.-West. Bundesanstalt für Geowissenschaften und Rohstoffe, Hannover.

Friedel, S., 1997. Hochauflösende Geoelektrik - Geoelektrische Tomographie. In: H. Beblo (Editor), Umweltgeophysik. Ernst & Sohn Verlag für Architektur und technische Wissenschaften GmbH, Berlin, pp. 131-151.

Fromm, K., 1978. Magnetostratigraphische Bestimmungen im Rhein-Main-Gebiet, Report, Niedersächsisches Landesamt für Bodenforschung, Hannover.

Garcia-Castellanos, D., Cloetingh, S. and Van Balen, R.T., 2000. Modelling the Middle Pleistocene uplift in the Ardennes-Rhenish Massif: thermo-mechanical weakening under the Eifel? Global and Planetary Change, 27: 39-52.

Gercek, H., 2007. Poisson's ratio values for rocks. International Journal of Rock Mechanics & Mining Sciences, 44: 1-13.

Giamboni, M., Ustaszewski, K., Schmid, S.M., Schumacher, M.E. and Wetzel, A., 2004. Plio-Pleistocene transpressional reactivation of Paleozoic and Paleogene structures in the Rhine-Bresse transform zone (northern Switzerland and eastern France). Int. J. Earth Sci. (Geol. Rundsch.), 93: 207-223.

Giamboni, M., Wetzel, A. and Schneider, B., 2005. Geomorphic Response of Alluvial Rivers to Active Tectonics: Example from the Southern Rhinegraben. Aust. J. Earth Sci. (Mitt. Österr. Geol. Gesell.), 97: 24-37.

Glatthaar, D., 1976. Die Entwicklung der Oberflächenformen im östlichen Rheinischen Schiefergebirge zwischen Lahn und Ruhr während des Tertiärs. Z. Geomorph. N.F. Suppl., 24: 79-87.

Gölke, M. and Coblentz, D., 1996. Origins of the European regional stress field. Tectonophysics, 266: 11-24.

Grimm, K., 2002. Tertiär, Mainzer Becken. In: German Stratigraphic Commission (Editor), Stratigraphic Table of Germany 2002.

Grimm, M.C., Hottenrott, M. and German Stratigraphic Commission, 2002. Tertiär, Oberrheingraben. In: German Stratigraphic Commission (Editor), Stratigraphic Table of Germany 2002.

Groshong, R.H., 1996. Construction and validation of extensional cross sections using lost area and strain, with application to the Rhine Graben. In: P.G. Buchanan and D.A. Nieuwland (Editors), Modern developments in structural interpretation. Geol. Soc. Spec. Publ., pp. 79-87.

Grünthal, G. and Stromeyer, D., 1992. The recent crustal stress field in Central Europe:

trajectories and finite element modeling. Journal Geophysical Research, 97(B8): 11,805-11,820.

Grünthal, G. and Wahlström, R., 2003. An M_w based earthquake catalogue for central, northern and northwestern Europe using a hierarchy of magnitude conversions. Journal of Seismology, 7: 507-531.

GÜK300, 1989. Geologische Übersichtskarte von Hessen 1:300000. Hessisches Landesamt für Bodenforschung, Wiesbaden.

GÜK500, 1998. Geologische Übersichtskarte von Baden-Württemberg 1:500000. Landesamt für Geologie, Rohstoffe und Bergbau Baden-Württemberg, Freiburg i.Br.

Gutenberg, R. and Richter, C.F., 1944. Frequency of earthquakes in California. Bull. Seism. Soc. Am., 34: 185-188.

Hack, J.T., 1973. Stream profile analysis and stream-gradient index. U.S. Geological Survey Journal of Research, 1: 421-429.

Hagedorn, E.-M., 2004. Sedimentpetrographie und Lithofazies der jungtertiären und quartären Sedimente im Oberrheingebiet. PhD thesis, Universität zu Köln, Köln, 310 pp.

Hägele, U. and Wohlenberg, J., 1970. Recent Investigations on the Seismicity of the Rheingraben Rift System. In: J.H. Illies and S. Mueller (Editors), Graben Problems, Stuttgart, pp. 167-170.

Haimberger, R., Hoppe, A. and Schäfer, A., 2005. High-resolution seismic survey on the Rhine River in the northern Upper Rhine Graben. Int. J. Earth Sci. (Geol. Rundsch.), 94(4): 657 - 668.

Heinemann, B., Troschke, B. and Tenzer, H., 1992. Hydraulic investigation and stress evaluations at the HDR test site Urach III, Germany, Geothermal Resources Council Transactions, Davis, CA, pp. 425-431.

Heitele, H., 1971. Gutachten über die geologischen und bodenmechanischen Verhältnisse im Bereich der Hochwasserrückhaltebecken Westhofen und Osthofen, Archive Geologisches Landesamt Rheinland-Pfalz GLA 6215, 32 (Westhofen), Mainz.

Henk, A., 1993. Subsidenz und Tektonik des Saar-Nahe-Beckens (SW-Deutschland). Geologische Rundschau, 82: 3-19.

Hergert, T., Buchmann, T., Eckert, A., Peters, G., Müller, B. and Heidbach, O., submitted. How to define an initial in-situ stress state for 3D numerical models? International Journal of Rock Mechanics & Mining Sciences.

Hinzen, K.G., 2003. Stress field in the Northern Rhine area, Central Europe, from earthquake fault plane solutions. Tectonophysics, 377: 325-356.

Höhl, G., Dörrer, I. and Schweinfurth, W., 1983. Erläuterungen zur Geomorphologischen Karte 1:25000 der Bundesrepublik Deutschland, GMK 25, Blatt 12, 6714 Edenkoben. Geo Center, Berlin.

Holbrook, J. and Schumm, S.A., 1999. Geomorphic and sedimentary response of rivers to tectonic deformation: a brief review and critique of a tool for recognizing subtle epeirogenic deformation in modern and ancient settings. Tectonophysics, 305: 287-306.

Hoselmann, C., 1996. Der Hauptterrassen-Komplex am unteren Mittelrhein. Z. dt. geol. Ges., 147(4): 481-497.

Houtgast, R.F., Van Balen, R.T. and Kasse, C., 2005. Late Quaternary tectonic evolution of the Feldbiss Fault (Roer Valley Rift System, the Netherlands) based on trenching, and its potential relation to glacial unloading. Quaternary Science Reviews, 24: 489-508.

Houtgast, R.F., Van Balen, R.T., Kasse, C. and Vandenberghe, J., 2003. Late Quatenary tectonic evolution and postseismic near surface fault displacements along the Geleen Fault (Feldbiss Fault Zone - Roer Valley Rift System, the Netherlands), based on trenching. Netherlands J. Geosciences, 82(2): 177-196.

Illies, J.H., 1967. Development and Tectonic Pattern of the Rhinegraben. Abh. geol. L.-Amt Baden-Württ., 6, The Rhinegraben Progress Report 1967: 7-9.

Illies, J.H., 1974a. Taphrogenesis and Plate Tectonics. In: K. Fuchs and J.H. Illies (Editors), Approaches to Taphrogenesis. Schweitzerbart'sche Verlagsbuchhandlung, Stuttgart, pp. 433-460.

Illies, J.H., 1974b. Taphrogenesis, Introductory Remarks. In: K. Fuchs and J.H. Illies (Editors), Approaches to Taphrogenesis. Schweitzerbart'sche Verlagsbuchhandlung, Stuttgart, pp. 1-13.

Illies, J.H., 1975. Intraplate tectonics in stable Europe as related to plate tectonics in the Alpine system. Geologische Rundschau, 64: 677-699.

Illies, J.H., 1977. Ancient and recent rifting in the Rhinegraben. Geologie en Mijnbouw, 56: 329-350.

Illies, J.H., Baumann, H. and Hoffers, B., 1981. Stress pattern and strain release in the Alpine Foreland. Tectonophysics, 71: 157-172.

Illies, J.H. and Fuchs, K., 1974. Approaches to Taphrogenesis. Approaches to Taphrogenesis, 8. E. Schweizerbart'sche Verlagsbuchhandlung (Nägele u. Obermiller), Stuttgart, 460 pp.

Illies, J.H. and Greiner, G., 1976. Regionales stress-Feld und Neotektonik in Mitteleuropa. Oberrhein. geol. Abh., 25: 1-40.

Illies, J.H. and Greiner, G., 1978. Rhinegraben and the Alpine system. Bull. Geol. Soc. America, 89: 770-782.

Illies, J.H., Prodehl, C., Schmincke, H.-U. and Semmel, A., 1979. The Quaternary uplift of the Rhenish Shield in Germany. Tectonophysics, 61: 197-225.

Ishimoto, M. and Iida, K., 1939. Observations of earthquakes registered with the microseismograph constructed recently. Bull. Earthquake Res. Inst., 17: 443-478.

Jackson, J., 1987. Active normal faulting and crustal extensino. In: M. Coward, J. Dewey and P. Hancock (Editors), Continental Extensional Tectonics. Blackwell, London, pp. 3-18.

Jaeger, J.C., 1969. Elasticity fracture and flow. Methuen & Co. Ltd., London, 268 pp.

Jost, M.L. and Hermann, R.B., 1989. A Student's Guide to and Review of Moment Tensors. Seismological Research Letters, 60: 37-57.

Jungbluth, F.A., 1918. Die Terrassen des Rheins von Andernach bis Bonn. Verhandlungen des naturhistorischen Vereins der preußischen Rheinlande und Westfalens, 73: 1-103.

Kaiser, D., 1999. Re-evaluation of the largest earthquakes in the northern Upper Rhinegraben (Germany) since the year 858, International Union of Geodesy and Geophysics XXII General Assembly, Birmingham.

Kaiser, E., 1903. Die Ausbildung des Rhein-Tales zwischen Neuwieder Becken und Bonn-

Cölner Bucht, Verhandlungen des 14. Deutschen Geographentages zu Cöln, pp. 206-215.

Kandler, O., 1970. Untersuchungen zur quartären Entwicklung des Rheintales zwischen Mainz/Wiesbaden und Bingen/Rüdesheim. Mainzer geogr.Stud., 3: 9-90.

Kárník, K. and Klíma, K., 1993. Magnitude-frequency distribution in the European-Mediterranean earthquake regions. Tectonophysics, 220(1-4): 309-323.

Keller, E.A. and Pinter, N., 1999. Active tectonics: earthquakes, uplift, and landscape. Prentice-Hall, New Jersey, 338 pp.

Klaer, W., 1977. Grundzüge der Naturlandschaftsentwicklung von Rheinhessen. Mainzer geogr.Stud., 11: 211-225.

Klee, G. and Rummel, F., 1999. Stress regime in the Rhinegraben basement and in the surrounding tectonic units. Bulletin d'Hydrogéologie, 17: 135-142.

Krantz, R.W., 1991. Measurements of friction coefficients and cohesion for faulting and fault reactivation in laboratory models using sand and sand mixtures. Tectonophysics, 188: 203-207.

Krauter, E. and Steingötter, K., 1983. Die Hangstabilitätskarte des linksrheinischen Mainzer Beckens. Geol. Jb., C34: 2-33.

Kreemer, C. and Holt, W.E., 2001. A No-net-rotation Model of Present Day Surface Motion. Geophysical Research Letters, 28: 4407-4410.

Lahner, L. and Toloczyki, M., 2004. Geowissenschaftliche Karte der Bundesrepublik Deutschland 1:2000000 - Geologie und Verkehr, GK2000. Bundesanstalt für Geowissenschaften und Rohstoffe, Hannover.

Lampe, C., 2001. The effects of hydrothermal fluid flow on the temperature history of the northern Upper Rhine Graben: Numerical simulation studies. Kölner Forum für Geologie und Paläontologie, 8: 1-126.

Larroque, J.M., Etchecopar, A. and Philip, H., 1987. Evidence for the permutation of stresses $\sigma 1$ and $\sigma 2$ in the Alpine foreland: the example of the Rhine graben. Tectonophysics, 144: 315-322.

Larroque, J.M. and Laurent, P., 1988. Evolution of the stress field pattern in the southern of the Rhine Graben from the Eocene to the present. Tectonophysics, 148: 41-58.

Laubscher, H., 2001. Plate interactions at the southern end of the Rhine graben. Tectonophysics, 343(1-2): 1-19.

Lehmann, K., Klostermann, J. and Pelzing, R., 2001. Paleoseismological Investigations at the Rurrand Fault, Lower Rhine Embayment. Netherlands J. Geosciences, 80: 139-154.

Lemeille, F., Cushing, M., Carbon, D., Grellet, B., Bitterli, T., Flehoc, C. and Innocent, C., 1999. Co-seismic ruptures and deformations recorded by speleothems in the epicentral zone of the Basel earthquake. Geodinamica Acta, 12(3-4): 179-191.

Leser, H., 1967. Beobachtungen und Studien zur quartären Landschaftsentwicklung des Pfrimmgebietes (Südrheinhessen). PhD Thesis, Universität Bonn, Bonn, 442 pp.

Leser, H., 1969. Landeskundlicher Führer durch Rheinhessen, 5. Gebrüder Borntraeger, Stuttgart, 253 pp.

Leydecker, G., 2005a. Erdbebenkatalog für die Bundesrepublik Deutschland mit Randgebieten für die Jahre 800 - 2004, Bundesanstalt für Geowissenschaften und Rohstoffe BGR,

Hannover.

Leydecker, G., 2005b. Projekt Gorleben - Standsicherheit Nachbetriebsphase: Seismische Gefährdung. Teilprojekt Ingenieurseismologie - Abschlussbericht -, Bundesanstalt für Geowissenschaften und Rohstoffe, Hannover.

Leydecker, G. and Harjes, H.-P., 1978. Seismische Kriterien zur Standortauswahl kerntechnischer Anlagen in der Bundesrepublik Deutschland. Abschlußbericht BMFT RS 170, Bundesanstalt für Geowissenschaften und Rohstoffe, Hannover.

Liniger, H., 1964. Beziehungen zwischen Pliozän und Jurafaltung in der Ajoie. Eclogae geol. Helv., 58: 215-229.

Lippert, W., 1979. Erstellung von Formeln für die Magnitudenbestimmung von Nahbeben sowie Seismizität der Jahre 1971-1978 im Bereich des Oberrheingrabens. Diploma thesis Thesis, University of Karlsruhe, Karlsruhe, 198 pp.

Lisle, R. and Srivastava, D.C., 2004. Test of the frictional reactivation theory for faults and validity of fault-slip analysis. Geology, 32(7): 569-572.

Liu, L. and Zoback, M.D., 1992. The Effect of Topography on the State of Stress in the Crust: Application to the Site of the Cajon Pass Scientific Drilling Project. Journal Geophysical Research, 97(B4): 5095-5108.

Loke, M.H., 2001. Electrical imaging surveys for environmental and engineering studies. A practical guide to 2-D and 3-D surveys, www.geoelectrical.com.

Lopes Cardozo, G.G.O., 2004. 3-D geophysical imaging and tectonic modelling of the active tectonics of the Upper Rhine Graben Region. PhD thesis, Vrije Universiteit, Amsterdam, 163 pp.

Lopes Cardozo, G.G.O. and Behrmann, J.H., 2006. Kinematic analysis of the Upper Rhine Graben boundary fault system. Journal of Structural Geology, 28: 1028-1039.

Lopes Cardozo, G.G.O., Edel, J.B. and Granet, M., 2005. Detection of active crustal structures in the Upper Rhine Graben using local earthquake tomography, gravimetry and reflection seismics. Quaternary Science Reviews, 24: 337-344.

Mackin, J.H., 1948. Concept of the graded stream. Geological Society of America Bulletin, 59: 463-512.

Manuskriptkarte, 1914. Geologische Manuskriptkarte GK25, 1:25000 Blatt 6515 Dürkheim Ost, Kartensammlung des Geologischen Landesamtes Rheinland-Pfalz, Mainz.

Marple, R.T. and Talwani, P., 1993. Evidence of possible tectonic upwarping along the South Carolina coastal plain from an examination of river morphology and elevation data. Geology, 21(7): 651-654.

Matthess, G., 1956. Beiträge zur geologischen Spezialkartierung des Blattes Alzey. Diploma thesis Thesis, TH Darmstadt, Darmstadt, 100 pp.

Mayer-Rosa, D. and Baer, M., 1992. Earthquake Catalogue of Switzerland, Swiss Federal Institute of Technology Zurich, Swiss Seismological Service, Zurich.

Mayer-Rosa, D. and Cadiot, B., 1979. A review of the 1356 Basel earthquake: basic data. Tectonophysics, 53: 325-333.

Mayer, G., Mai, P.M., Plenefisch, T., Echtler, H., Lüschen, E., Wehrle, V., Müller, B., Bonjer, K.-P., Prodehl, C. and Fuchs, K., 1997. The deep crust of the Southern Rhine Graben:

reflectivity and seismicity as images of dynamic processes. Tectonophysics, 275: 15-40.

McClay, K.R. and Ellis, P.G., 1987. Geometries of extensional fault systems developed in model experiments. Geology, 15: 341-344.

McGarr, A. and Gay, N.C., 1978. State of stress in the earth's crust. Ann. Rev. Earth Planet. Sci., 6: 405-436.

Meghraoui, M., Camelbeeck, T., Vanneste, K., Brondeel, M. and Jongmans, A.G., 2000. Active faulting and paleoseismology along the Bree fault, Lower Rhine graben, Belgium. J. Geophys. Res., 105(B6): 13809-13841.

Meghraoui, M., Delouis, B., Ferry, M., Giardini, D., Huggenberger, P., Spottke, I. and Granet, M., 2001. Active Normal Faulting in the Upper Rhine Graben and Paleoseismic Identification of the 1356 Basel Earthquake. Science, 293: 2070-2073.

Meier, L. and Eisbacher, G.H., 1991. Crustel kinematics and deep structure of the northern Rhine graben, Germany. Tectonics, 10(3): 621-630.

Merrits, D. and Vincent, K.R., 1989. Geomorphic response on coastal streams to low, intermediate and high rates of uplift, Mendocino Triple Junction region, Northern California. Geological Society of America Bulletin, 110: 1373-1388.

Meyer, W., Albers, H.J., Berners, H.P., von Gehlen, K., Glatthaar, D., Löhnertz, W., Pfeffer, K.H., Schnütgen, A., Wienecke, K. and Zakosek, H., 1983. Pre-Quaternary Uplift in the Central Part of the Rhenish Massif. In: K. Fuchs, K. von Gehlen, H. Mälzer, H. Murawski and A. Semmel (Editors), Plateau Uplift: The Rhenish Shield - A Case History. Springer Verlag, Heidelberg, pp. 39-46.

Meyer, W. and Stets, J., 1996. Das Rheintal zwischen Bingen und Bonn. Sammlung geologischer Führer, Bd. 89. Borntraeger, Stuttgart, 386 pp.

Meyer, W. and Stets, J., 1998. Junge Tektonik im Rheinischen Schiefergebirge und ihre Quantifizierung. Z. dt. geol. Ges., 149: 359-379.

Meyer, W. and Stets, J., 2002. Pleistocene to Recent Tectonics in the Rhenish Massif (Germany). Netherlands J. Geosciences, 81: 217-221.

Michon, L., Van Balen, R.T., Merle, O. and Pagnier, H., 2003. The Cenozoic evolution of the Roer Valley Rift System integrated at a European scale. Tectonophysics, 367: 101-126.

Mogi, K., 1967. Earthquakes and fractures. Tectonophysics, 5(1): 35-55.

Monecke, K., Anselmetti, F.S., Becker, A., Sturm, M. and Giardini, D., 2004. The record of historic earthquakes in lake sediments of Central Switzerland. Tectonophysics, 394(1-2): 21-40.

Monninger, R., 1985. Neotektonische Bewegungsmechanismen im mittleren Oberrheingraben. PhD hesis, Universität Karlsruhe, Karlsruhe, 219 pp.

Morris, A., Ferrill, D.A. and Henderson, D.B., 1996. Slip-tendency analysis and fault reactivation. Geology, 24(3): 275-278.

Müller, B., Zoback, M.L., Fuchs, K., Mastin, L., Gregersen, S., Pavoni, N., Stephansson, O. and Ljunggren, C., 1992. Regional patterns of tectonic stress in Europe. Journal Geophysical Research, 97(B8): 11783-11803.

Nocquet, J.-M. and Calais, E., 2004. Geodetic Measurements of Crustal Deformation in the Western Mediterranean and Europe. Pure appl. geophys., 161: 661-681.

Obermeier, S.F., 1996a. Use of liquefaction-induced features for paleoseismic analysis - An overview of how seismic liquefaction features can be distinguished from other features and how their regional distribution and properties of source sediment can be used to infer the location and strength of Holocene paleo-earthquakes. Engineering Geology, 44: 1-76.

Obermeier, S.F., 1996b. Using Liquefaction-Induced Features For Paleoseismic Analysis. In: J.P. McCalpin (Editor), Paleoseismology, San Diego, pp. 331-397.

Oestreich, K., 1909. Studien zur Oberflächengestaltung des Rheinischen Schiefergebirges. PM, 3: 57-63.

Ouchi, S., 1985. Response of alluvial rivers to slow active tectonic movement. Geological Society of America Bulletin, 96(4): 504-515.

PALEOSIS, 2000. Evaluation of the potential for large earthquakes in regions of present-day low seismicity activity in Europe. Final report, project no. ENV4-CT97-0578, Directorate-General XII for Science, Research and Development, Brussels.

Penck, A., 1910. Versuch einer Klimaklassifikation auf physiographischer Grundlage, Sitzungsberichte der Preussischen Akademie der Wissenschaften, Physikalisch-Mathematische Classe, Berlin, pp. 236-246.

Peters, G., Buchmann, T.J., Connolly, P., van Balen, R., Wenzel, F., Cloetingh, S.A.P.L. (2005). Interplay between tectonic, fluvial and erosional processes along the Western Border Fault of the northern Upper Rhine Graben, Germany. Tectonophysics 406: 39-66.

Peters, G., van Balen, R.T. (2007). Pleistocene tectonics inferred from fluvial terraces of the northern Upper Rhine Graben, Germany. Tectonophysics 430: 41-65.

Peters, G., van Balen, R.T. (2007). Tectonic geomorphology of the northern Upper Rhine Graben, Germany. Global Planetary Change 58: 310-334.

Petersen, M.D., Cramer, C.H., Reichle, M.S., Frankel, A.D. and Hanks, T.C., 2000. Discrepancy between earthquake rates implied by historic earthquakes and a consensus geologic source model for California. Bull. Seism. Soc. Am., 90: 1117-1132.

Pflug, R., 1982. Bau und Entwicklung des Oberrheingrabens. Erträge der Forschung, 184. Wissenschaftliche Buchgesellschaft, Darmstadt, 145 pp.

Plenefisch, T. and Bonjer, K.-P., 1997. The stress field in the Rhine Graben area inferred from earthquake focal mechanisms and estimation of frictional parameters. Tectonophysics, 275: 71-97.

Pottgiesser, T. and Halle, M., 2004. Entwicklung einer (Abschnitts-)Typologie für den natürlichen Rheinstrom - Endbericht -. Bericht Nr. 146d, Internationale Kommission zum Schutz des Rheins (IKSR), i.A. Umweltbüro Essen, Essen.

Prodehl, C., Mueller, S. and Haak, V., 1995. The European Cenozoic rift system. In: P.E. Olsen (Editor), Developments in Tectonics - Continental Rifts: Evolution, Structure, Tectonics. Elsevier Sci., New York, pp. 133-212.

Quitzow, H.W., 1974. Das Rheintal und seine Entstehung. Bestandsaufnahme und Versuch einer Synthese. Centenaire de la Société Géologique de Belgique, L'évolution quaternaire des bassins fluviaux de la mer du nor méridionale: 53-104.

Ramsay, J.G. and Lisle, R.J., 2000. The Techniques of Modern Structural Geology, Volume 3: Applications of Continuum Mechanics in Structural Geology. Academic Press, London,

1061 pp.

Reches, Z., 1987. Determination of the tectonic stress tensor from slip along faults that obey the Coulomb yield condition. Tectonics, 6(6): 849-861.

Reinecker, J., Heidbach, O., Tingay, M., Connolly, P. and Müller, B., 2004. The 2004 release of the World Stress Map (available online at www.world-stress-map.org).

Richardson, R.M., 1992. Ridge forces, absolute plate motions, and the intraplate stress field. Journal Geophysical Research, 97(B8): 11,738-11,748.

Richardson, R.M., Solomon, S.C. and Sleep, N.H., 1979. Tectonic stress in the plates. Reviews of Geophysics, 17(5): 981-1019.

Ritter, J.R.R., Jordan, M., Christensen, U.R. and Achauer, U., 2001. A mantle plume below the Eifel volcanic fields, Germany. Earth and Planetary Science Letters, 186: 7-14.

Rogall, M. and Schmitt, S.-O., 2005. Hangstabilitätskarte des linksrheinischen Mainzer Beckens 1:50000. Landesamt für Geologie und Bergbau, Mainz.

Roll, A., 1979. Versuch einer Volumenbilanz des Oberrheintalgrabens und seiner Schultern. Geol. Jb., A(52): 3-82.

Rothausen, K. and Sonne, V., 1984. Mainzer Becken. Sammlung geologischer Führer, 79. Gebrüder Borntraeger, Berlin - Stuttgart, 203 pp.

Rózsa, S., Heck, B., Mayer, B., Seitz, K., Westerhaus, M. and Zippelt, K., 2005. Determination of displacements in the upper Rhine graben Area from GPS and leveling data. International Journal of Earth Sciences, 94(4): 538-549.

Rudloff, A. and Leydecker, G., 2002. Ableitung von empirischen Beziehungen zwischen der Lokalbebenmagnitude und makroseismischen Parametern - Ergebnisbericht, BGR, Hannover.

Rummel, F. and Baumgärtner, J., 1991. Hydraulic Fracturing Stress Measurements in the GPK1 Borehole, Soultz-sous-Forêts. Geothermal Science and Technology, 3(1-4): 119-148.

Rummel, F., Möhring-Erdmann, G. and Baumgärtner, J., 1986. Stress constraints and hydrofracturing stress data for the continental crust. Pure and Applied Geophysics, 124(4-5): 875-895.

Ruszkiczay-Rüdiger, Z., Dunai, T.J., Bada, G., Fodor, L. and Horváth, E., 2005a. Middle to late Pleistocene uplift rate of the Hungarian Mountain Range at the Danube Bend, (Pannonian Basin) using in situ produced ^{3}He. Tectonophysics, 410: 173-187.

Ruszkiczay-Rüdiger, Z., Fodor, L., Bada, G., Leél-Össy, S., Horváth, E. and Dunai, T.J., 2005b. Quantification of Quaternary vertical movements in the central Pannonian Basin: A review of chronologic data along the Danube River, Hungary. Tectonophysics, 410: 157-172.

Scharpff, H.-J., 1977. Geologische Karte von Hessen 1:25000, Erläuterungen, Blatt 6316 Worms. Hessisches Landesamt für Bodenforschung, Wiesbaden.

Scheer, H.D., 1978. Gliederung und Aufbau der Niederterrassen von Rhein und Main im nördlichen Oberrheingraben. Geol. Jb. Hessen, 106: 273-289.

Scheer, H.D. and Wourtsakis, A., 1986. Bodenkarte von Rheinland-Pfalz 1:25000, Erläuterungen Blatt 6415 Grünstadt-Ost. Geologisches Landesamt Rheinland-Pfalz, Mainz.

Scheffer, F. and Schachtschabel, P., 1998. Lehrbuch der Bodenkunde, Stuttgart, 494 pp.

Schmedes, J., Hainzl, S., Reamer, S.-K., Scherbaum, F. and Hinzen, K.-G., 2005. Moment release in the Lower Rhine Embayment, Germany: seismological perspective of the deformation process. Geophys. J. Int., 160: 901-909.

Schnarrenberger, K., 1985. Geologische Karte 1:25000 von Baden-Württemberg, Blatt 6917 Weingarten (Baden). Geologisches Landesamt Baden-Württemberg, Stuttgart.

Schneider, E.F. and Schneider, H., 1975. Synsedimentäre Bruchtektonik im Pleistozän des Oberrheintal-Grabens zwischen Speyer, Worms, Hardt und Odenwald. Münster. Forsch. Geol. Paläont., 36: 81-126.

Scholz, C.H., 1968. The frequency-magnitude relation of microfracturing in rock and its relation to earthquakes. Bull. Seism. Soc. Am., 58: 399-415.

Scholz, C.H., 2002. The Mechanics of Earthquakes and Faulting. Cambridge University Press, Cambridge, 471 pp.

Schraft, A., 1979. Das Neogen am Westrand des Oberrheingrabens bei Oppenheim. Oberrhein. geol. Abh., 28: 29-39.

Schumacher, M.E., 2002. Upper Rhine Graben: Role of preexisting structures during rift evolution. Tectonics, 21(1): 10.1029/2001TC900022.

Schumm, S.A., 1986. Alluvial river response to active tectonics, Active Tectonics. National Academy Press, Washington, D.C., pp. 80-95.

Schumm, S.A., Dumont, J.F. and Holbrook, J.M., 2000. Active Tectonics and Alluvial Rivers. Cambridge University Press, 276 pp.

Schwarz, E., 1974. Levelling Results at the Northern End of the Rhinegraben. In: K. Fuchs and J.H. Illies (Editors), Approaches to Taphrogenesis, Stuttgart, pp. 261-268.

Schwarz, E., 1976. Präzisionsnivellement und rezente Krustenbewegung dargestellt am nördlichen Oberrheingraben. Z. f. Vermessungswesen, 14-25.

Schwarz, J., Beinersdorf, S., Golbs, C., Ahorner, L. and Meidow, H., 2006. Erdbebenkatalog für Deutschland und angrenzende Gebiete - erweiterter Ahorner-Katalog (EKDAG2006). Auszug mit Schadenbeben der Intensität ≥ VI, Bauhaus-Universität Weimar, Erdbebenzentrum, Köln/Weimar.

Schwarz, M., 2006. Evolution und Struktur des Oberrheingrabens quantitative Einblicke mit Hilfe dreidimensionaler thermomechanischer Modellrechnungen. PhD thesis, University of Freiburg, Freiburg i. Br., 337 pp.

Schwarz, M. and Henk, A., 2005. Evolution and structure of the Upper Rhine Graben: insights from three-dimensional thermomechanical modelling. Int. J. Earth Sci. (Geol. Rundsch.), 94: 732–750.

Semmel, A., 1968. Studien über den Verlauf jungpleistozäner Formung in Hessen. Frankfurter geograph. Hefte, 45: 133.

Semmel, A., 1969. Das Quartär. In: E. Kümmerle and A. Semmel (Editors), Geologische Karte von Hessen 1:25.000, Erläuterungen, Blatt Nr. 5916 Hochheim a. Main. Hessisches Landsamt für Bodenforschung, Wiesbaden, pp. 51-99.

Semmel, A., 1978. Untersuchungen zur quartären Tektonik am Taunus-Südrand. Geol. Jb. Hessen, 106: 291-302.

Semmel, A., 1979. Geomorphological criteria for recent tectonic - A discussion of examples from the northern Upper Rhine area. Allg. Vermess. Nachr., 86: 370-374.

Semmel, A., 1983. The Early Pleistocene Terraces of the Upper Middle Rhine and Its Southern Foreland - Questions Concerning Their Tectonic Interpretation. In: J.H. Illies et al. (Editors), Plateau Uplift: The Rhenish Massif - A Case History. Springer Verlag, Heidelberg, pp. 49-54.

Semmel, A., 1986. Angewandte konventionelle Geomorphologie - Beispiele aus Mitteleuropa und Afrika. Frankfurter Geowissenschaftliche Arbeiten, Band 6, Serie D. Fachbereich für Geowissenschaften der Johann W.G. Universität Frankfurt, Frankfurt a.M., 1-83;108-115 pp.

Semmel, A., 1991. Neotectonics and geomorphology in the Rhenish Massif and the Hessian Basin. Tectonophysics, 195: 291-297.

Semmel, A. and Fromm, K., 1976. Ergebnisse paläomagnetischer Untersuchungen an quartären Sedimenten des Rhein-Main-Gebiets. Eiszeitalter u. Gegenwart, 27: 18-25.

Shackleton, N.J., 1987. Oxygen isotopes, ice volume and sea level. Quaternary Science Reviews, 6: 183 - 190.

Sheorey, P.R., 1994. A Theory for In Situ Stresses in Isotropic and Transversely Isotropic Rock. Int. J. Rock Mech. Min. Sci & Geomech. Abstr., 31(1): 23-34.

Sibson, R.H., 1974. Frictional constraints on thrust, wrench and normal faults. Nature, 249: 542-544.

Sibson, R.H. and Xie, G.Y., 1998. Dip range for intracontinental reverse fault ruptures: Truth not stranger than friction? Bull. Seismol. Soc. Amer., 88: 1014-1022.

Silva, P.G., Goy, J.L., Zazo, C. and Bardaji, T., 2003. Fault-generated mountain fronts in southeast Spain: geomorphologic assessment of tectonic and seismic activity. Geomorphology, 50(1-3): 203-225.

Simpson, R.W., 1997. Quantifying Anderson´s fault types. Journal Geophysical Research, 102(B8): 17,909-17,919.

Sissingh, W., 1998. Comparative Tertiary stratigraphy of the Rhine Graben, Bresse Graben and Molasse Basin: correlation of Alpine Foreland events. Tectonophysics, 300: 249-284.

Sittler, C. and Sonne, V., 1971. Vorkommen und Verbreitung eozäner Ablagerungen im nördlichen Mainzer Becken. N. Jb. Geol. Paläont. Mh.: 372-384.

Sonne, V., 1956. Ein Beitrag zur geologischen Spezialkarierung des Blattes Alzey. Diploma thesis, TH Darmstadt, Darmstadt, 65 pp.

Sonne, V., 1969. Die Entwicklung des Alzey-Niersteiner Horstes seit Beginn des Tertiärs. Jber. Mitt. oberrhein. geol. Ver., N.F., 51: 81-86.

Sonne, V., 1972. Geologische Karte von Rheinland-Pfalz 1:25000, Erläuterungen zu Blatt 6115 Undenheim. Geologisches Landesamt Rheinland-Pfalz, Mainz.

Sonne, V., 1977. Tiefenlinienplan des Talbodens der Rhein-Niederterrasse zwischen Budenheim bei Mainz und Bingen-Kempten. Mz. Naturw. Arch., 16: 83-90.

Sonne, V., 1989. Geologische Karte von Rheinland-Pfalz 1:25000, Erläuterungen, Blatt 6015 Mainz. Geologisches Landesamt Rheinland-Pfalz, Mainz.

Sonne, V. and Weiler, H., 1984. Die detritischen altertiären (oligozänen) Faunen- und Florenelemente in den Sedimenten des Meerfelder Maares. Cour. Forsch.-Inst.

Senckenberg, 65: 87-95.

Stäblein, G., 1968. Reliefgenerationen der Vorderpfalz - Geomorphologische Untersuchungen im Oberrheingraben zwischen Rhein und Pfälzer Wald. Würzburger geogr. Arbeiten, 23: 1-183.

Stahmer, G., 1979. Geologie und Stratigraphie des Raumes Grünstadt - Monsheim / Pfalz. Diploma thesis, University of Heidelberg, Heidelberg, 92 pp.

Stapf, K.R.G., 1988. Zur Tektonik des westlichen Rheingrabens zwischen Nierstein am Rhein und Wissembourg (Elsaß). Jber. Mitt. oberrhein. geol. Ver., N.F., 70: 399-410.

Stein, S. and Wysession, M., 2003. An introduction to seismology, earthquakes, and earth structure. Blackwell Publishing, 512 pp.

Steuer, A., 1911. Geologische Karte des Großherzogtums Hessen im Maßstabe 1:25000, Erläuterungen, Blatt Oppenheim. Hessisches Landesamt für Bodenforschung, Wiesbaden.

Straub, E.W., 1962. Die Erdöl- und Erdlagerstätten in Hessen und Rheinhessen. Abh. geol. Landesamt Baden-Württemberg, 4: 123-136.

Streit, J.E. and Hillis, R.R., 2004. Estimating fault stability and sustainable fluid pressures for underground storage of CO_2 in porous rock. Energy, 29: 1445-1456.

Tenzer, H., Budeus, P. and Schellschmidt, R., 1992. Fracture analyses in hot dry rock drillholes at Soultz and Urach by Borehole televiewer measurements, Geothermal Resources Council Transactions, Davis, CA, pp. 317-321.

Tesauro, M., Hollenstein, C., Egli, R., Geiger, A. and Kahle, H.-G., 2005. Continuous GPS and broad-scale deformation across the Rhine Graben and the Alps. International Journal of Earth Sciences, 94(4): 525 - 537.

Tietze, R., Neeb, I., Walgenwitz, F. and Maget, P., 1979. Geothermische Synthese des Oberrheingrabens (Bestandsaufnahme) - Synthèse géothermique du fossè rhénan supérieur (Etat des connaissances), Geologisches Landesamt Baden-Württemberg, Service Géologique Régional Alsace, Freiburg, Strasbourg.

Tsulukidze, P.P., Burchuladze, S.V. and Mikashvili, Y.N., 1973. Effect of the age of concrete used in hydraulic structures on its strength and deformative properties in compression. Power Technology and Engineering, 7(11): 1040-1043.

Turcotte, D.L. and Schubert, G., 2002. Geodynamics. Cambridge University Press, 456 pp.

Tuttle, M.P., 2001. The use of liquefaction features in paleoseismology: Lessons learned in the New Madrid seismic zone, central United States. J. Seismology, 5(3): 361-380.

Twiss, R.J. and Moores, E.M., 1992. Structural Geology. Freeman Co., New York, 532 pp.

Urbancic, T.I., Trifu, C.-I., Long, J.M. and Young, P.G., 1992. Space-time correlations of b values with stress release. Pure and Applied Geophysics, 139: 449-462.

Van Balen, R.T., Houtgast, R.F., Van der Wateren, F.M., Vandenberghe, J. and Bogaart, P.W., 2000. Sediment budget and tectonic evolution of the Meuse catchment in the Ardennes and the Roer Valley Rift System. Global and Planetary Change, 27: 113-129.

Van Balen, R.T., Vandenberghe, J. and Kasse, K. (Editors), 2003. Quaternary Science Reviews. Fluvial response to rapid environmental change, 22 (20), 2053-2235 pp.

Van den Berg, M.W., Vanneste, K., Dost, B., Lokhorst, A., van Eijk, M. and Verbeeck, K., 2002. Paleoseismological investigations along the Peel Boundary Fault: geological setting, site selection and trenching results. Netherlands J. Geosciences, 81: 39-60.

Vanneste, K., Meghraoui, M. and Camelbeeck, T., 1999. Late Quaternary earthquake-related soft-sediment deformation along the Belgian portion of the Feldbiss Fault, Lower Rhine Graben system. Tectonophysics, 309: 57-79.

Vanneste, K. and Verbeeck, K., 2001. Paleoseismological analysis of the Rurrand fault near Jülich, Roer Valley graben, Germany: Coseismic or aseismic faulting history? Netherlands J. Geosciences, 80: 155-169.

Vanneste, K., Verbeeck, K., Camelbeeck, T., Paulissen, E., Meghraoui, M., Renardy, F., Jongmans, D. and Frechen, M., 2001. Surface-rupturing history of the Bree fault scarp, Roer Valley graben: Evidence for six events since the late Pleistocene. J. Seismology, 5(3): 329-359.

Vendeville, B., Cobbold, P.R., Davy, P., Brun, J.P. and Choukroune, P., 1987. Physical models of extensional tectonics at various scales. In: M.P. Coward, J. Dewey and P. Hancock (Editors), Continental Extensional Tectonics. Geological Society Special Publication, London, pp. 95-107.

Vigny, C., Chery, J., Duquesnoy, T., Jouanne, F., Ammann, J., Anzidei, M., Avouac, J.-P., Barlier, F., Bayer, R., Briole, P., Calais, E., Cotton, F., Duquenne, F., Feigl, K.L., Ferhat, G., Flouzat, M., Gamond, J.-F., Geiger, A., Harmel, A., Kasser, M., Laplanche, M., Le Pape, M., Martinod, J., Menard, G., Mayer, B., Ruegg, J.-C., Scheubel, J.-M., Scotti, O. and Vidal, G., 2002. GPS network monitors the Western Alps' deformation over a five-year period: 1993-1998. J. Geodesy, 76: 63-76.

Villemin, T., Alvarez, F. and Angelier, J., 1986. The Rhinegraben: Extension, Subsidence and Shoulder Uplift. Tectonophysics, 128: 47-59.

Wagner, W., 1930. Bemerkungen zur Tektonischen Skizze des westlichen Mainzer Beckens. Notizbl. Ver. Erdkde. und hess. geol. Landesamt, V(12): 185-188.

Wagner, W., 1931a. Die ältesten linksrheinischen Diluvialterrassen zwischen Oppenheim, Mainz und Bingen. Notizbl. d. Ver. f. Erdk. u. hess. geol. Landesamt, 14: 31-45.

Wagner, W., 1931b. Geologischen Karte von Hessen im Maßstabe 1:25000, Erläuterungen, Blatt Ober-Ingelheim. Hessischer Staatsverlag, Darmstadt.

Wagner, W., 1935. Geologischen Karte von Hessen im Maßstabe 1:25000, Blatt Wörrstadt. Hessischer Staatsverlag, Darmstadt.

Wagner, W., 1941a. Bodenversetzungen und Bergrutsche im Mainzer Becken. Geol. u. Bauwesen, 13: 17-33.

Wagner, W., 1941b. Geologisch-hydrologisches Gutachten im Gebiet von Westhofen - Gundersheim, Archive Geologisches Landesamt Rheinland-Pfalz GLA 6215, 34, Mainz.

Wagner, W., 1962. Der Rhein im Rheintalgraben und im Mainzer Becken - Eine Schiffsreise von Karlsruhe nach Bingen. Beitr. Rheinkunde, 14 (Die Entstehung des Rheintales vom Austritt des Flusses aus dem Bodensee bis zur Mündung): 22-34.

Wagner, W., 1972. Über Pleistozän und Holozän in Rheinhessen (Mainzer Becken). Mainzer geowiss. Mitt., 1: 192-197.

Wallace, R.E., 1951. Geometry of shearing stress and relation to faulting. Journal of Geology, 59: 118-130.

Weidenfeller, M. and Kärcher, T., 2004. Terrassen, Kieslager und Zwischenhorizonte - Neue Aspekte zur geologisch-hydrologischen Gliederung der quartären Sedimente im linksrheinischen Teil des nördlichen Oberrheingrabens. From Source to Delta, DEUQUA meeting, 30 August - 3 September 2004, Nijmegen, the Netherlands.

Weidenfeller, M. and Zöller, L., 1995. Mittelpleistozäne Tektonik in einer Löß-Paläoboden-Abfolge am westlichen Rand des Oberrheingrabens. Mainzer geowiss. Mitt., 24: 87-102.

Weiler, W., 1931. Die diluvialen Terrassen der Pfrimm mit einem Anhang über altdiluviale Säuger aus der Umgebung von Worms. Notizbl. d. Ver. f. Erdk., V.(13): 124-145.

Weiler, W., 1952. Pliozän und Diluvium im südlichen Rheinhessen I. Teil: Das Pliozän und seine organischen Einschlüsse. Notizbl. hess. L.-Amt Bodenforsch., 6(3): 147-170.

Wells, D.L. and Coppersmith, K.J., 1994. New Empirical Relationships among Magnitude, Rupture Length, Rupture Width, Rupture Area, and Surface Displacement. Bulletin of the Seismological Society of America, 84(4): 974-1002.

Wells, S.G., Bullard, T.F., Menges, C.M., Drake, P.G., Karas, P.A., Kelson, K.I., Ritter, J.B. and Wesling, J.R., 1988. Regional variations in tectonic geomorphology along a segmented convergent plate boundary pacific coast of Costa Rica. Geomorphology, 1(3): 239-265.

Wenzel, F., Brun, J.P. and ECORS-DEKORP team, 1991. A deep reflection seismic line across the Northern Rhine Graben. Earth and Planetary Science Letters, 104: 140-150.

Wenzel, F., Peters, G. and Connolly, P.T., 2004. Potenzial paläoseismologischer Erkenntnisse für die Quantifizierung der Standortsicherheit von Nuklearanlagen - Potential of paleoseismic investigation for quantification of the site stability of nuclear power plants, Geophysical Institute, University of Karlsruhe, Karlsruhe, 112 pp.

Wiemer, S. and Wyss, M., 1997. Mapping the frequency-magnitude distribution in asperities: an improved technique to calculate recurrence times? Journal Geophysical Research, 102: 15,115-15,128.

Wittmann, O., 1955. Bohnerz und präeozäne Landoberfläche im Markgräflerland. Jh. geol. Landesamt Baden-Württemberg, 1: 267-299.

Worum, G., 2004. Modelling of fault reactivation potential and quantification of inversion tectonics in the southern Netherlands. PhD thesis, Vrije Universiteit, Amsterdam, 152 pp.

Worum, G., van Wees, J.D., Bada, G., van Balen, R.T., Cloetingh, S.A.P.L. and Pagnier, H., 2004. Slip tendency analysis as a tool to constrain fault reactivation: A numerical approach applied to three-dimensional fault models in the Roer Valley Rift System (southeast Netherlands). J. Geophys. Res., 109(B02401): 10.1029-10.1044.

Zeis, S., Gajewski, D. and Prodehl, C., 1990. Crustal structure of southern Germany from seismic refraction data. Tectonophysics, 176: 59-86.

Zhao, S. and Müller, R.D., 2003. Three-dimensional finite-element modelling of the tectonic stress field in continental Australia (Chapter 6). Geological Society of Australia Special Publication, 22: 65-83.

Ziegler, P.A., 1992. European Cenozoic rift system. Tectonophysics, 208: 91-111.

Ziegler, P.A., 1994. Cenozoic rift system of western and central Europe: An overview. Geologie en Mijnbouw, 73: 99-127.

Ziegler, P.A., Bertotti, G. and Cloetingh, S., 2002. Dynamic processes controlling foreland development - the role of mechanical (de)coupling of orogenic wedges and foreland. EGU Special Publication, 1: 17-56.

Ziegler, P.A. and Dèzes, P., 2005. Evolution of the lithosphere in the area of the Rhine Rift System. Int. J. Earth Sci. (Geol. Rundsch.), 94: 594-614.

Ziegler, P.A., Van Wees, J.-D. and Cloetingh, S., 1995. Geodynamics of intraplate compressional deformation; the Alpine foreland and other examples. Tectonophysics, 252: 7-61.

Zienert, A., 1989. Geomorphological aspects of the Odenwald. In: F. Ahnert (Editor), Landforms and landform evolution of West Germany, pp. 199-210.

Zienkiewicz, O.C., Taylor, R.L. and Zhu, J.Z., 2005. The Finite Element Method. Its Basis and Fundamentals. Butterworth-Heinemann, 752 pp.

Zippelt, K. and Mälzer, H., 1981. Recent height changes in the central segment of the Rhinegraben and its adjacent shoulders. Tectonophysics, 73: 119-123.

Zoback, M.D. and Healy, J.H., 1984. Friction, faulting, and insitu stress. Annales Geophysicae, 2: 689-698.

Zoback, M.D. and Healy, J.H., 1992. In situ stress measurements to 3.5 km depth in the Cajon Pass scientific research borehole: implications for the mechanics of crustal faulting. Journal Geophysical Research, 97: 5039-5057.

Zoback, M.L., Zoback, M.D., Adams, J., Assumpção, M., Bell, S., Bergman, E.A., Blümling, P., Brereton, N.R., Denham, D., Ding, J., Fuchs, K., Gay, N., Gregersen, S., Gupta, H.K., Gvishiani, A., Jacob, K., Klein, R., Knoll, P., Magee, M., Mercier, J.L., Müller, M.C., Paquin, C., Rajendran, K., Stephansson, O., Suarez, G., Suter, M., Udias, A., Xu, Z.H. and Zhizhin, M., 1989. Global patterns of tectonic stress. Nature, 341: 291-298.